High-Mix
Low-Volume
Manufacturing

Hewlett-Packard Professional Books

Atchison	Object-Oriented Test & Measurement Software Development in C++
Blinn	Portable Shell Programming: An Extensive Collection of Bourne Shell Examples
Blommers	Practical Planning for Network Growth
Caruso	Power Programming in HP OpenView: Developing CMIS Applications
Cook	Building Enterprise Information Architectures
Costa	Planning and Designing High Speed Networks Using 100VG-AnyLAN, Second Edition
Crane	A Simplified Approach to Image Processing: Classical and Modern Techniques
Fernandez	Configuring the Common Desktop Environment
Fristrup	USENET: Netnews for Everyone
Fristrup	The Essential Web Surfer Survival Guide
Grady	Practical Software Metrics for Project Management and Process Improvement
Grosvenor, Ichiro, O'Brien	Mainframe Downsizing to Upsize Your Business: IT-Preneuring
Gunn	A Guide to NetWare® for UNIX®
Helsel	Graphical Programming: A Tutorial for HP VEE
Helsel	Visual Programming with HP VEE, Second Edition
Holman, Lund	Instant JavaScript
Kane	PA-RISC 2.0 Architecture
Knouse	Practical DCE Programming
Lee	The ISDN Consultant: A Stress-Free Guide to High-Speed Communications
Lewis	The Art & Science of Smalltalk
Lund	Integrating UNIX® and PC Network Operating Systems
Madell	Disk and File Management Tasks on HP-UX
Mahoney	High-Mix Low-Volume Manufacturing
Malan, Letsinger, Coleman	Object-Oriented Development at Work: Fusion in the Real World
McFarland	X Windows on the World: Developing Internationalized Software with X, Motif®, and CDE
McMinds/Whitty	Writing Your Own OSF/Motif Widgets
Norton, DiPasquale	Thread Time: The Multithreaded Programming Guide
Phaal	LAN Traffic Management
Pipkin	Halting the Hacker: A Practical Guide to Computer Security
Poniatowski	The HP-UX System Administrator's "How To" Book
Poniatowski	HP-UX 10.x System Administration "How To" Book
Poniatowski	Learning the HP-UX Operating System
Poniatowski	The Windows NT and HP-UX System Administrator's How-To Book
Ryan	Distributed Object Technology: Concepts and Applications
Thomas	Cable Television Proof-of-Performance: A Practical Guide to Cable TV Compliance Measurements Using a Spectrum Analyzer
Weygant	Clusters for High Availability: A Primer of HP-UX Solutions
Witte	Electronic Test Instruments
Yawn, Stachnick, Sellars	The Legacy Continues: Using the HP 3000 with HP-UX and Windows NT

High-Mix
Low-Volume
Manufacturing

R. Michael Mahoney

Hewlett-Packard Company

To join a Prentice Hall PTR Internet mailing list, point to
http://www.prenhall.com/register

Prentice Hall PTR
Upper Saddle River, New Jersey 07458
http://www.prenhall.com

Library of Congress Cataloging in Publication Data

Mahoney, R. Michael.
　　　High-mix low-volume manufacturing / R. Michael Mahoney.
　　　　　　p.　　cm -- (Hewlett-Packard Professional books)
　　　Includes bibliographical references and index.
　　　ISBN 0-13-255688-X (alk. paper)
　　　1. Production management.　2. Production planning.　3. Production scheduling.
　　　I. Title.　II. Series.
TS155.M328　　1997
658.5--dc21　　　　　　　　　　　　　　　　　　96-1613
　　　　　　　　　　　　　　　　　　　　　　　　　CIP

Editorial/Production Supervision: *Kathleen M. Caren*
Cover Design Director: *Jerry Votta*
Manufacturing Manager: *Alexis Heydt*
Marketing Manager: *Miles Williams*
Acquisitions Editor: *Karen Gettman*
Editorial Assistant: *Barbara Alfieri*
Book design: *Keith Rosenhagen, Graphic Relations, Loveland, CO*
Manager, Hewlett-Packard Press: *Patricia Pekary*

Published by Prentice-Hall PTR.
Prentice-Hall, Inc.
Upper Saddle River, New Jersey 07458

The publisher offers discounts on this book when ordered in bulk quantities.
For more information, contact:
　　　　　Corporate Sales Department
　　　　　Prentice Hall PTR
　　　　　1 Lake Street
　　　　　Upper Saddle River, NJ　07458
　　　　　Phone: 1-800-382-3419
　　　　　FAX: 201-236-7141
　　　　　E-mail: corpsales@prenhall.com

Printed in the United States of America
10　9　8　7　6　5　4

ISBN 0-13-255688-X

PRENTICE-HALL INTERNATIONAL (UK) LIMITED, *LONDON*
PRENTICE-HALL OF AUSTRALIA PTY. LIMITED, *SYDNEY*
PRENTICE-HALL CANADA INC., *TORONTO*
PRENTICE-HALL HISPANOAMERICANA, S.A., *MEXICO*
PRENTICE-HALL OF INDIA PRIVATE LIMITED, *NEW DELHI*
PRENTICE-HALL OF JAPAN, INC., *TOKYO*
PEARSON EDUCATION ASIA PTE. LTD., *SINGAPORE*
EDITORA PRENTICE-HALL DO BRASIL, LTDA., *RIO DE JANEIRO*

To my brother
Mark D. Mahoney

Table of Contents

Chapter 1: High-Mix Manufacturing Competitive Strategy

Chapter 2: Critical Manufacturing Fundamentals

Chapter 3: Cross-Functional Relationships

Chapter 4: Multiple Constraint Synchronization (MCS)

Chapter 5: Performance Measurement

List of Illustrations

Chapter 1

Chapter 2

Chapter 3

Chapter 4

Chapter 5

Foreword

This is an unusual book and a valuable contribution to stimulating improvements in manufacturing. New literature on operations management has been characterized by:

1. Texts for academic and industrial courses with selected contents and narrow scope.
2. Case studies of successful applications of popular concepts and techniques.
3. Explanations of "how to" implement now popular (not always new) concepts and/or techniques in specific environments.
4. Strong sales pitches (with new buzzwords) for their computer hardware, software, and education courses offered by sales personnel and nonpractitioner educators.

This book is none of the above. The author is an engineer on manufacturing development projects for a successful large corporation; his knowledge of the book's subject has a solid foundation in theory and practice, and he has only the book to sell. While the focus is on high-mix, low-volume manufacturing, probably the most difficult to manage well, there is much useful and valuable information for people in all other types of manufacturing.

The book covers the full spectrum of manufacturing-related topics from competitive strategies through performance measurements. Compacted into just over 200 pages, this is a reference book that belongs in the library of anyone interested in how manufacturing should and can be run. Throughout, it spotlights the importance of education to develop fully qualified people at all levels of manufacturing organizations.

Strategic planning is well covered, including the key players, criteria in customer-based versus competitor-based strategies, and leapfrogging not just catching competitors. Master planning for three major types of manufacturing—build-to-stock, build-to-order, and assemble-to-order—has two components: master production scheduling driving the formal computer-based planning system and final assembly scheduling driving execution. Current strategies of continuous improvement, restructuring, reengineering, and reinventing are compared and contrasted.

Also included are ways to evaluate product sensitivity to selling price changes and the pitfalls of using conventional accounting methods in costing and profit contribution calculations. Useful information is included on Activity Based Costing and methods of evaluating alternative investment decisions for more flexibility and controlling costs of improved responsiveness.

Highlights of manufacturing fundamentals include Just-In-Time to synchronize supplier deliveries with customer needs, Kanban as a replacement for system-based data to provide simple visible controls of material flow, Theory of Constraints to improve output of bottleneck facilities having borderline or inadequate capacity and random interruptions in output, and MRP II to permit management involvement in integrating all manufacturing plans with its supply chain firms. Scheduling techniques and their importance in operations get adequate coverage.

The key role of managing inventory is covered including the dangers to cost control and customer service of focusing only on improving return on investment. Techniques are illustrated for simulating the effects on inventory, costs, and on-time deliveries of changing lot sizes.

Agile manufacturing, increasing flexibility, and improving responsiveness to customer demand are linked to capacity management. A chapter on strategic capacity planning shows it as a fundamental requirement of sound planning to ensure producing enough in total to meet all demands. Ways are shown to minimize capacity imbalance among work centers to lessen pressures on constraints. Also included are techniques for modeling operations, synchronizing multiple moving constraints, lot sequencing, transfer lot sizing, and buffer inventory provision.

Improving quality is an important topic. Various ways are shown to analyze different types of errors affecting process quality together with advanced techniques for improving effectiveness of in-process testing. These include the proper location of repair facilities.

Ending on a high note, the characteristics of world-class companies are presented. This book is not an easy read, but the effort to understand the author's contributions to improving manufacturing will be very worthwhile.

George W. Plossl

Preface

This book is intended for practitioners and academics interested in obtaining a sound understanding of the underlying principles of high-mix, low-volume manufacturing. This book opens the door to the many applications of manufacturing planning and control system techniques. Although there is a special emphasis on complex high-mix, low-volume manufacturing environments, the principles and techniques presented are universally applicable.

This book provides a holistic framework in which to organize and unify the development of manufacturing capabilities that will be in alignment with the organizational strategy developed at the highest level of an organization. This book is directed toward the specific goal of developing the capacity of a manufacturing manager or academic to creatively design manufacturing capabilities that will result in competitive advantage as opposed to competitive parity.

The central enigma of high-mix, low-volume manufacturing is unlocked in a way that often gets overlooked by corporate strategists. To be an effective manufacturing manager, an understanding of manufacturing must extend beyond the framework of only simplified and idealized models. The manufacturing manager must thoroughly understand the interrelationships involved in the global manufacturing system problem. Additionally, he or she must know, and understand, how to implement the various strategies available.

This book plays an important educational role. It will assist the academic and practitioner in the effort to obtain a clear and understandable picture of how a high-mix, low-volume manufacturing environment can and should be managed. This book will guarantee a solution to the high-mix, low-volume manufacturing problem but cannot guarantee the solution will be followed. Solutions cannot be effectively implemented without having the patience and stomach to examine the problems confronting a manufacturing organization. Problems are not solved because they are not clearly understood. This book performs a thorough diagnosis of the many problems encountered by a manufacturer in a way that will move management toward action. Additionally, managers will be less likely to undermine themselves by rationalizing problems in a way that leaves the solution outside of their control.

This book will enable an organization to build a bridge across the chasm of organizational politics. Organizational politics is about the exercise of power. Managerial resistance to new ideas is often a result of defending their own adequacy. This book circumvents the managerial political problem as well as the fears associated with loss of control.

This book can be used by an industrial engineering or second year MBA course on manufacturing planning and control systems. Concepts are developed from the simple to the complex. In this regard, the most appropriate approach to using this book is in strict table of contents sequence.

The opportunity to single-mindedly devote myself to the production of this book was based on five years of work. Although the bibliography of 166 works represents a fraction of the total influence on my thinking, they are the most important. These works, combined with the American Production and Inventory Control Society (APICS) curricula for Certification in Production and Inventory Management (CPIM), encapsulate the body of profound knowledge that I have come to understand and cherish.

I have had the good fortune to debate a broad range of manufacturing topics with Greg Beam (Hewlett-Packard Manufacturing Development Engineer), Greg Larsen (Hewlett-Packard Statistician), and Scott Stever (Hewlett-Packard Research and Development Manager). In every sense, this book is a reflection of those debates. I am also in gratitude to Sara L. Beckman, Ph.D. (Co-director, Management of Technology Program, University of California), Cihan H. Dagli, Ph.D. (Department of Engineering Management, University of Missouri), George W. Plossl (G. W. Plossl & Company, Inc.), Rob Saffer (Hewlett-Packard Research and Development Engineer), Greg Thoman (Hewlett-Packard Manufacturing Development Engineer), and Wendy Wyckaert (Hewlett-Packard Industrial Engineer) for their consideration, review, and recommendations for improving the manuscript.

R. Michael Mahoney
Longmont, Colorado
Winter 1997

CHAPTER 1

High-Mix Manufacturing Competitive Strategy

1.1 Building a Competitive Advantage

When building a competitive advantage, the principal underlying concern is to avoid doing the same thing on the same battlefield as your competition. You must achieve a competitive advantage by gaining a relative superiority through methods and measures the competitor will find difficult to follow. Once this is achieved, you must extend that superiority as far as possible. History can be leveraged to expand the various courses of action and facilitate intelligent choices among the various options. During the endeavor to achieve competitive advantage, the thinking process required of the manufacturing manager is no less than that required of the military officer drawing up a strategic plan for a battalion engagement. You must focus on those key factors that will bring success.

When developing a competitive strategy, three key players must be taken into account: the company, the customer, and the competitor. To be effective, representatives from all critical manufacturing functions must comprise the strategic planning group. The strategic planning group must deploy functional expertise that includes the thorough and detailed knowledge required to know *how* to implement the desired strategy. The advantage(s) achieved must be persistently deployed over competitors who are unable to close the gap. The competitors will therefore be losers in the battle for customers.

A strategic success formula does not exist. Creativity and innovation are essential ingredients in overcoming a competitor. While creative strategies are extremely hard to duplicate, we must also crystallize the boundaries defining our probability of success. The boundaries are a function of reality, timing, and resources. The strategy must consistently succeed in satisfying customers more effectively than any competitor's strategy. Profit will follow.

1.2 Strategic Degrees of Freedom

The concept of strategic degrees of freedom defines the amount of freedom for strategic moves of a particular success factor. While the theoretical possibilities for improvements affecting high-mix, low-volume manufacturing environments are innumerable, not all of them will produce equally significant results. The main objective behind assessing the degrees of strategic freedom is to avoid the waste of time and money that is bound to occur if management fails to determine in advance the best direction for improvement. A company must perform a ruthlessly objective assessment of its internal weaknesses and strengths prior to determining which direction for improvement the company should embark upon. Direction and resources should be concentrated to give the best possible chance for obtaining competitive advantage.

Competitive advantages are not created in conventional reciprocal head-on competition. If your competitor volume-focuses his factory, you volume-focus yours; if the competitor cuts the price, you cut yours. Such tactics never work very well for very long. Approaches such as these are competitor-based strategies and not customer-based. In effect, they will allow a company's profitability to be controlled by its competitors. A thorough study of all possibilities will often produce an abundance of profitable customer-based strategies that can provide a competitive advantage rather than competitive parity. The framework for such a study must address four key criteria through a process called **SWOT** analysis:

- The **S**trengths of the company relative to its customers and competitors.
- The **W**eaknesses of the company relative to its customers and competitors.
- Creation of new marketplace **O**pportunities through market segmentation or fragmentation.
- **T**hreats to the company resulting from known or pending strategic or tactical moves by competitors.

Creativity, innovation, and risk taking are necessary to facilitate the learning necessary to secure a competitive advantage. Success must be summoned. Success will not come unplanned. Strategic thinking must break out of the limited scope of vision that entraps deer on the highway. Flexible thinking is imperative and is demonstrated by daily use of imagination and by continual training in logical manufacturing thought processes.

1.3 Evolution to High-Mix Manufacturing

Any credible historical perspective of manufacturing competitive strategy must focus on Japan. Japan is responsible for several competitive breakthroughs in manufacturing since the mid-1960s. The historical perspective is important not in terms of a detailed understanding of past strategies but in understanding what is no longer viable and what will be effective in today's and tomorrow's competitive global marketplace and why. The age-old adage that those who do not study history are condemned to repeat it has never applied more resolutely than to manufacturing strategy. See Fig. 1.3.1.

Let us begin at the conclusion of World War II. Japanese strategy was focused on low labor cost. At this time Japan manufactured cheap products with infamously poor quality. In the late 1950s Dr. W. Edwards Deming arrived and revolutionized quality in Japan. Quality has to this day remained a key competitive cost reduction strategy that attracts and keeps customers. The key to obtaining competitive advantage through quality centers is the prevention of internal defects. The Japanese refer to this as Poka-Yoke. Quality, of course, is ultimately gauged by the customer. The associated reduction in costs that occurs in contrast to a competitor's is what creates increased margins and competitive advantage. Amazingly enough, to this day many American companies believe that obtaining the prestigious Deming Quality Award (Japan) or the Malcolm Baldrige National Quality Award (United States) is cost-prohibitive. This is an ancient and obsolete understanding of quality that can literally put a company out of business if its competitors embrace quality as a competitive weapon and are competing on

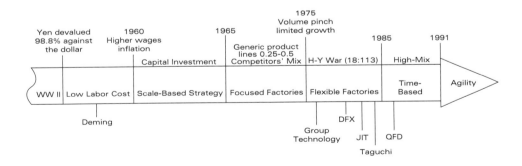

Figure 1.3.1 Evolution to high-mix manufacturing.

price. The Japanese have priced many American companies out of business while remaining profitable. Beginning around 1960, Japan shifted gears to respond to the increased demand for their quality products. Large capital investments were made in their manufacturing infrastructure to exploit the consequence of economies of scale using volume-focused factories. At about the mid-1960s the Japanese began to introduce variety to their customer base. Varieties were approximately 25 to 50 percent of their competitors' mixes. Generic product lines were maintained to continue their exploitation of economies of scale through volume-focused factories to maintain low cost, high quality production. This strategy began to falter. Customers were demanding more choices and the Japanese were experiencing limited growth by not being able to respond. In order to grow, the Japanese realized that higher volumes would only be achieved by increasing the mix of their product offering. The Japanese would soon learn that introducing variety into a volume-focused repetitive manufacturing environment is a flawed business decision. Nowhere is this more vividly demonstrated than in the Honda-Yamaha war, or the H-Y war as it has come to be known. Yamaha publicly announced in Japan that it would invest in the infrastructure necessary to become the largest motorcycle producer in the world. Honda responded decisively to this competitive threat. In the span of about 18 months, Honda introduced 113 new products. Customers flocked to Honda, and Yamaha was left financially devastated. Yamaha succumbed to Honda and the war ended. Both companies suffered greatly. Inventories and costs were exorbitant. The process-related improvements and core competencies required to economically produce a high variety of products were nonexistent. The H-Y war has, once and for all, demonstrated that volume-focusing a high-mix manufacturing environment is a business decision destined to fail. Despite the history of failure, there are many companies today that are volume-focusing high-mix manufacturing environments.

United States:
- GM
- Ford
- Chrysler

Europe:
- Volkswagen
- Mercedes-Benz
- Renault
- PSA
- BMW
- Fiat
- Volvo
- Rover
- Porsche
- SAAB (GM)
- Opel (GM)

Japan:
- Toyota
- Nissan
- Honda
- Mitsubishi
- Mazda
- Isuzu
- Suzuki
- Hyundai
- Daihatsu
- Fugi Hvy Truck

Figure 1.3.2 Automotive companies—OEMs.

United States:
- The Limited
- Federal Express
- Wilson Art
- Domino's Pizza
- McDonald's
- Benetton

Japan:
- Toyota
- Sony
- NEC
- Toshiba
- Matsushita
- Hitachi
- Honda
- Hino

Figure 1.3.3 Companies using time-based strategies.

Realizing that growth would occur with increased variety in their product offering, the Japanese embarked on a path that brought economies of scale to increased product mix. During the period from about 1975 to 1985, Group Technology, Design for Manufacturability and Assembly, Just-In-Time, and Taguchi's Design of Experiments were effectively leveraged to reduce costs and improve delivery performance. Once again the Japanese sustained world-class performance levels and growth. Despite its beginnings in the 1970s, the Toyota Just-In-Time manufacturing cost reduction strategy remains a source of competitive advantage for manufacturing companies that are recognized as world-class producers today.

It is important to understand that the major thrust behind the evolution of Japanese manufacturing strategy is the result of competition that exists within Japan and not the external world competitive environment. Using automobiles as an example, there are 10 automotive manufacturers in Japan and all are viable today (Fig. 1.3.2).

In the true spirit of fierce competition, the Japanese began to shift gears once again. Beginning around 1985, time-based competitive strategy began to unfold. The Japanese are now managing time instead of money. The central focus is to create a system in which value-added time as a proportion of total time is maximized throughout the entire value delivery chain. It is interesting to note the differences between American and Japanese companies using time-based strategies today. The contrast of the American service industry focus versus the product manufacturing focus in Japan is quite remarkable (Fig. 1.3.3). Time-based competition represents an evolutionary change to the Toyota Just-In-Time production system. In the course of time-based competition, Quality Function Deployment (QFD) was invented by the Japanese. QFD is essentially a disciplined system for translating customer requirements into company requirements all the way through product development to the factory floor. Realizing that infinite variety does not yield infinite growth, the Japanese are now strategically, and with surgical precision, analytically determining the varieties that customers want and are benchmarking their competitors to ensure they are designing superior products that offer a competitive advantage. Speed kills, but in business speed is fatal to the competition. By managing time throughout the value delivery chain the way less adept competitors manage costs, the Japanese are addressing what customers care about. To customers, what matters the most is the total time required to deliver a low-cost, high-quality product or service. Manufacturing is just one part of this total, but it must be optimized first.

1.4 Strategic Cost Drivers

In order for a company to grow in good times and bad, low-cost variety and quick response times are fundamental competitive differentiators. Increased variety increases costs due to complexity. Variety costs include inventory, material handling, setup, test, and other overhead costs. The cost of complexity increases in proportion to the logarithm of variety increases. The rate of increase is approximately 20 to 35 percent for every doubling of complexity.[†] Many costs are sensitive to volume, and economies of scale are generally process related. Economies of scale improve as volumes increase. The economies of scale improve approximately 15 to 25 percent for every doubling of volume.[†] Thus, total manufacturing cost is the sum of scale and complexity-driven costs (Fig. 1.4.1).

To achieve competitive advantage in a high-mix manufacturing environment, the costs of complexity must be lower than the competitors' costs. Achieving such changes results in a more flexible and agile manufacturing environment. The point of minimum total cost will occur at higher volumes

† *Source: The Boston Consulting Group*

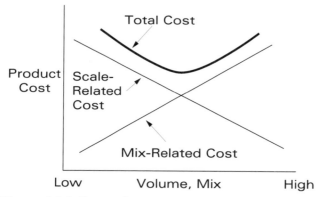

Figure 1.4.1 Strategic cost drivers.

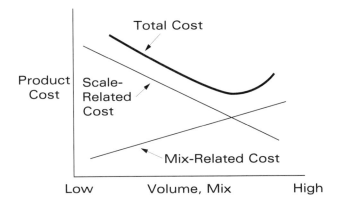

Figure 1.4.2 Strategic cost drivers.

and variety. The total cost is reduced both absolutely and on margin (Fig. 1.4.2) as compared to traditional lean manufacturing environments.

Consider the situation where a high product-variety manufacturer wants to volume-focus production operations. To do this, the mix on half of the production lines is reduced to a subset of products that represent the greatest volumes. The lower volume products removed from the volume-focused production lines are added to the remaining production lines. We have thus created factories within a factory where half are volume-focused and the other half are variety-focused. This strategy is a zero-sum solution. While it is true that the volume-focused factory will achieve economies of scale due to the increased volumes and reduction in mix, it is also true that the variety-focused factory will exhibit increased costs due to increased complexity. It is important to realize that the increases in mix-related costs grow at a faster rate than the cost savings achieved through economies of scale. Given a constant or increasing variety of products, the competitor who reduces the costs associated with complexity will always outperform the competitor who focuses on volume. Volume-focusing a high-mix manufacturing production environment is not a core competency.

1.5 Agile Manufacturing

In 1991 an industry task force was convened at the Iacocca Institute, Lehigh University. The members of this task force laid the foundations of what is known as agile competition. Rather than offering the customer a plethora of different options and models from which to choose, the customer works with the producer to arrive at solutions to the customer's specific problem. The responsibility of choice moves to the producer. Information and services

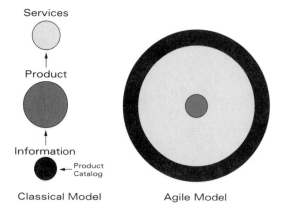

Figure 1.5.1 Enhancing customer value through agility.

become parts of the product sold to customers. See Fig. 1.5.1.

Companies that are agile are able to fragment markets in a way that undermines the competitiveness of the lean mass production systems that exist today. Products are now information and services loaded. Agility is the successor of mass production manufacturing environments. To support agility, alliances are formed among companies across worldwide geographic expanses and are integrated to provide a custom customer solution. Such a cooperative alliance is called a virtual company. The only constraint in this environment will be the constantly increasing variety of products. Product design will become more modular to facilitate the ease of configuration required to economically satisfy the requirements of a particular customer solution. The economics of production will no longer define competitive advantage in terms of being the low-cost supplier. The economics are primarily determined by the customer's perception of the solution's value and the solution provided will be priced based on this value. A company can command a large premium for providing information and services-loaded solutions. Responsiveness is the competitive differentiator in this environment and those that only compete on price will cease to exist. Agile competition requires significantly different techniques than those required for lean Just-In-Time manufacturing environments. The Japanese are finding themselves very unknowledgeable about this new competitive reality. Many manufacturing companies are backing into agility unknowingly. Only those companies that strategically seek agility and develop high-mix manufacturing core competencies that offer a competitive advantage will win the battle for customers.

1.6 Historical Heuristics

Heuristics (derived from the Greek word heuriskein, meaning to discover) are rules which will aid in the discovery of one or more solutions to a specific problem. Heuristic methods are based on inductive inferences. In the search for acceptable high-mix manufacturing solutions, a particular heuristic can be followed based on historical experience and the learning derived from the experience of using the rule. While heuristic problem solving involves the use of currently accepted rules, it also involves a search for superior rules to replace them. When searching for acceptable solutions to manufacturing-related problems, the use of trial-and-error methods is expensive and inefficient and should be eschewed. High-mix manufacturing is a very large complex problem, and an attempt to solve this problem using a direct analytical solution is impractical. A large class of production scheduling problems fall into this category. Rather than mutilate a manufacturing problem until it conforms to a model, the best approach is to modify the solution procedure to fit the problem so as to obtain a directed search of the solution space. Complex high-mix manufacturing problems have no suitable analytic solution available and do not have optimal solutions in the strict sense.

Heuristic problem solving is associated with satisficing as opposed to optimizing behavior. For example, goal satisficing is likely to be most prevalent at higher levels of management in manufacturing organizations. Most evident are the trade-offs made where goal conflicts arise. One approach to solving high-mix manufacturing problems is to study historical methods, each possessing certain preferred characteristics in varying degrees. The decision maker can now make trade-off decisions between conflicting characteristics and does not have to make arbitrary assumptions. Heuristics play a large role in reducing the search for an acceptable decision. The complex game of chess is a case in point. Chess books abound with such rules. We must avoid high-mix operational decisions of which it can be said that the surgery was successful but the patient died. A high-mix manufacturing operation cannot survive many such deaths. What we can learn from history is essential to efficiently and effectively unlocking the complexity of high-mix manufacturing and learn how to create a competitive advantage.

1.7 The Competitiveness Sigmoid Curve

The sigmoid curve, or "S" curve, can be used to describe a broad spectrum of time-based events such as profits, revenue, and product life cycles. When applied to competitiveness, some interesting realities emerge. If we were to

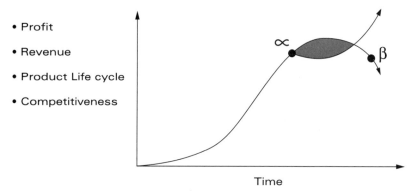

Figure 1.7.1 The sigmoid curve.

ask manufacturing employees where they were on the sigmoid curve, point α or point β, many would select point β (Fig. 1.7.1). When developing a competitive strategy we must assume that we are near the apex of the curve rather than on the declining portion of the curve. A company on the decline will probably not have the necessary resources available to finance new competitive strategies. Companies on the decline are losing money and downsizing valuable people resources, not because of poor productivity, but due to a flawed business strategy. Usually only large companies with significant financial reserves can withstand such a competitive onslaught. Several American companies have experienced this decline as a result of the superiority of their Japanese competitors. The steel and semiconductor industries are cases in point. When limited growth occurs, the time to change strategy is now. The startling truth is that time compression is affecting the rate at which companies will have to change in the future. The more rapidly you can gain relative superiority over a competitor, the greater the chance you have to improve on that advantage. If your competitor does not respond quickly enough, it may put its very existence at risk. In business competition there are only winners and losers.

CHAPTER 2

Critical Manufacturing
Fundamentals

2.1 Just-In-Time Manufacturing (JIT)

Mr. Taiichi Ohno, V.P. at Toyota Motor Company, was not oblivious to the Honda-Yamaha war and the high inventories and costs that are incurred when high-mix manufacturing is coupled with volume-focused factories. Mr. Ohno first introduced a computer-based system that sends orders to vendors, lists them in the order of production, gives sufficient advance notice (based on the vendor's lead time requirements), and a production plan specifying type, quantity, delivery time, and order of delivery.

Kanban is a signaling tool used to support on-time deliveries to meet the automotive assembly schedule. The key to JIT is that suppliers are synchronizing their delivery schedule with their customer's production schedule, thus eliminating work in process (WIP). The suppliers hold component stocks right up to the time they are needed and deliver them according to plan. It is important to understand that the suppliers' production schedules are not the same as those of their customers, only the delivery schedules are. What's more important is that JIT is relatively robust to vendor production lead times. The delivery time is the critical path. Delivery time can be a source of competitive advantage when suppliers are in close geographical proximity to their customers. JIT allows a company to strip inventories from its production system in a way that contributes to a low-cost position, and accomplish this without damaging its ability to give good customer service. Higher labor productivity may be achieved by fostering a situation that embraces workers as partners and gains their expertise in the workplace. Additionally, management in a JIT manufacturing environment considers decisions about the work force, inventory control and quality as strategic decisions and not operating decisions.

The "heart" of JIT is the reduction of setup time. Reductions in setup time can be leveraged to develop schedules that yield clear competitive advantages. To demonstrate this, consider a simplified manufacturing system. Two printed circuit board manufacturers (company, competitor) are suppliers to their respective final assembly operations. Both companies will produce identical printed circuit boards (PCB X, PCB Y, PCB Z) and one of each is required to produce one complete electronic instrument. Both production systems are identical in every respect except for the use of different scheduling methods and differing setup times. The company's setup time is one minute and the competitor's is 30 minutes. The processing time for each supplier is equal to one time unit for each PCB and there are orders placed for 30 instruments. The processing time for both final assembly operations is the same. If the company produces to the schedule: X, Y, Z, X, Y, Z, ..., the setup time is one minute per unit. If the competitor wants to deliver all of its PCBs at the same time as the company, it is going to have to amortize the adverse effects of setup time. To accomplish this, the competitor will have to produce each PCB type in a batch of 30; 30 minute setup time/30 units = 1 minute setup time per unit. Let us now consider the consequences of these two different scheduling strategies. The company will supply enough PCBs to facilitate the completion of one electronic instrument every six time units:

setup + PCB X + setup + PCB Y + setup + PCB Z + ...

The competitor will not respond to the customer with enough PCBs to complete the first electronic instrument until a total of 151 time units have elapsed:

setup + 30 PCB X + setup + 30 PCB Y + setup + PCB Z + ...

The company is over 25 times as responsive as the competitor *(151/6)*. Responsiveness is defined as the minimum time required to produce at least one of each PCB type while delivery time is defined as the total time required to deliver the complete order for 30 instruments. The delivery time is the same for both companies. When responsiveness is a competitive differentiator, the company will decisively defeat the competitor in the marketplace.

The inventory investment for both companies is dramatically different. The company will have PCB components delivered to satisfy a lot size of one, while the competitor is going to have PCB components delivered in lots of 30. The setup time difference coupled with a proper schedule is ultimately responsible for the company's competitive advantage over its competitor.

Let us now modify the example so that the company and competitor are supplying PCBs to three separate customers such that 30 PCB Xs, 30 PCB Ys, and 30 PCB Zs are ordered by customer A, customer B, and customer C respectively. The competitor's poor response time will likely result in the loss of customer C. Most likely, customer C will take its business to the company.

Suboptimal quality will have an adverse impact on the delivery performance for the company and the competitor. Consider the situation where a particular PCB X component part is being supplied by a vendor whose production process is out of control to the extent that all of the component parts produced are out of electrical specification. If we assume that the defective parts can only be detected at the final assembly testing stage, the delivery performance for the company and competitor will be significantly different. The company will detect the component problem after only one assembled PCB X is delivered to the final assembly test process. The competitor will deliver the total lot of 30 completely assembled PCB Xs to test before detecting the faulty component. The company will detect the faulty component 30 times *(60/2)* as soon as the competitor. Faster feedback cycles for the company will enable corrective actions to be taken thereby not all of the remaining PCB Xs will be loaded with the defective component. This results in reduced repair costs and lessens the impact on delivery date performance. It is too late for the competitor. Relative to the competitor, the company has a competitive advantage in delivery performance. Despite the clear advantages that can be achieved through setup time reduction and scheduling, many manufacturers today persist in executing large batches through their production facilities. In the special case where processing times for multiple production process steps are equal, an operations schedule can be generated so that the manufacturing lead time required to satisfy differing demand quantities will be the same. Such a case is typical of automotive manufacturers. Consider the following demands placed on a car manufacturer (see Table 2.1.1):

Table 2.1.1

Order Type	Quantity	% Total
Model W	2,000	20
Model X	1,000	10
Model Y	4,000	40
Model Z	3,000	30
Total	10,000	100

A repetitive sequence can be generated that will smooth the demand and provide excellent responsiveness and customer delivery service.

The repetitive sequence is generated as follows:

1	2	3	4	5	6	7	8	9	10	% Demand
Y			Y			Y		Y		40 Model Y
	Z			Z					Z	30 Model Z
		W					W			20 Model W
					X					10 Model X

Y	Z	W	Y	Z	X	Y	W	Y	Z

Final Schedule I

If the sequence Y, Z, W, Y, Z, X, Y, W, Y, Z is repeated 1,000 times, the total demand for all models will be satisfied during the 1,000th iteration. Note that the number of setups required is maximized, and the actual demand is equal to the scheduled demand. A less responsive schedule such as Final Schedule II would be required if, for example, the setup time for Final Schedule II were two times the setup time for Final Schedule I:

Y	Y	Z	Z	W	W	Y	Y	Z	Z	X	X	Y	Y	W	W	Y	Y	Z	Z

Final Schedule II

Final Schedule II is half as responsive as Final Schedule I and would have to be repeated 500 times to complete the total demand for all items in the same makespan as Final Schedule I. In terms of responsiveness, a schedule such as:

Y	Y	Y	Y	Y	Y	Y	Y	Z	Z	Z	Z	Z	Z	W	W	W	W	X	X

Final Schedule III

that is nothing more than a regrouping of Final Schedule II, would exhibit reduced customer responsiveness. It is critical to understand that a company's

2,000(W)	1,000(X)	4,000(Y)	3,000(Z)

Final schedule IV

setup time requirements offer a competitive advantage only when proper scheduling is performed. The company's responsiveness is ultimately determined by the production schedule itself. If you based a schedule on the incoming customer order demand quantities from Table 2.1.1, the consequences would be disastrous. There are many problems with this scheduling philosophy:

- Minimum responsiveness
- High cost of quality
- Poor delivery performance
- Poor inventory position.

Final Schedule IV would likely put a car manufacturer out of business in today's highly competitive automotive industry. Despite the problems associated with adopting the philosophy of setting lot sizes based on total incoming order demand quantities, many companies persistently adopt this philosophy.

The typical strategy to address the poor performance exhibited by large batch production scheduling methods such as Final Schedule IV is to increase capacity. A 100 percent increase in capacity will not enable Final Schedule IV to be as responsive as Final Schedule III. For example, if we assume that the time to process each unit for Final Schedule III is one time unit, then the average time to process each unit for Final Schedule IV will be half a time unit. Final Schedule III will respond to the customer by shipping the mix of products at a maximum of every 19 time units. For Final Schedule IV, the first Z unit will not be available until the 3,500.5 time unit. This is highly likely to result in the loss of customers who ordered the Z models. Final Schedule III is an astounding 18,324 percent more responsive *{[(3500.5-19)/19]100}* than Final Schedule IV, despite doubling the capacity for the competitor using Final Schedule IV. Increasing capacity to solve the responsiveness problem, plainly and simply, does not work as effectively as proper scheduling techniques. In fact, additional capital investments to increase capacity will further damage the competitive position of a company by increasing overall costs with minimal improvements in customer responsiveness. This explains in clear terms why the Japanese were able to significantly outperform American automotive manufacturers for as long as they did and do it with less manufacturing space. This is a powerful testimony to the power of scheduling. When the specter of insufficient capacity occurs, an even worse strategy is to outsource a portion of the manufacturing base. This results in an unnecessary direct loss of important revenue. Once outsourcing starts, it is difficult to stop. This is particularly true if the vendor's prices are competitive with your costs. Management may begin to question why they are in the manufacturing business at all, and before you know it, the entire manufacturing base is gone.

All of the aforementioned production schedules are response mechanisms to actual customer demand and are equal to exact customer demand. Given equal setup times and processing times, the competitor who does the better job of scheduling will obtain a competitive advantage through improved customer responsiveness. Scheduling is therefore a customer-based strategy. Final Schedule IV represents a company-based strategy that amortizes setup time to the point that it is negligible. This strategy obtains better equipment utilization at the expense of the customer and many cost accounting systems encourage this sort of behavior. Localized process performance metrics based on financial data are irrelevant to optimizing overall manufacturing performance. Relevance has been lost in many cost accounting approaches used today. JIT scheduling is most appropriately used in manufacturing environments where process steps have relatively similar processing times. For JIT manufacturing environments *line balancing* is the condition where process steps have equal processing times for the variety of products produced. For high-mix manufacturing environments where dissimilar products use the same manufacturing resources, the processing times at each process step for the mix of products produced will most likely differ. Line balancing will not be achievable.

In a JIT manufacturing environment it is important to have cross-trained flexible workers to maintain production flow during absenteeism as a result of production department meetings, illness, etc. Workers should not operate cross-functionally as a matter of course to respond to variability encountered due to batch production methods. Workers are used cross-functionally based on clearly defined circumstances in a JIT manufacturing environment. Attempting to use worker flexibility to respond to fluctuations resulting from batch production methods will result in certain chaos. The subset of the work force that has the best aptitude for being cross-trained and flexible must be determined. This will help to ensure that the production plan is achieved. Should excessive unplanned absenteeism or quality problems occur, overtime should be used. Every effort must be made to achieve the *daily* production schedule. Manufacturing superiority using JIT is the result of implementation on the manufacturing floor and is not driven by cultural, work ethic, or personnel policy differences between the United States and Japan. JIT starts with a drive to reduce production lot sizes through reduced setup times and focuses on relentless cost reduction through the elimination of waste.

Continuous process improvement is the ethos for a company that embraces JIT as a competitive weapon. Small lot sizes permit key linkages between subsequent process steps that facilitate dramatic improvements in internal quality levels. The interdependence of production operations is significantly enhanced. When lots are large, workers are prone to sift out the good items versus the defective ones, when there is pressure to meet promised customer delivery dates. The reverse psychology is true for small lot production. In small lot production, downstream problems are produced immediately. The need to avoid further defects is obvious. This naturally leads to teamwork. Workers are now making improvements because of the production system and not in spite of it. Quality circles came into existence in Japan after JIT. The responsibilities engendered by JIT production are self-generated. Employee participation and empowerment are results of the production situation. Attempts to empower the work force and obtain continuous quality improvement without a sound underlying system of support are doomed to failure. Dr. Edwards W. Deming[37,38] made this abundantly clear when he asserted that 85 to 92 percent of all quality problems are assignable to the system. He went on to say that only management can change the system, and therefore the primary responsibility for quality improvement is management's. We can now understand the wisdom of his words. Many manufacturers today use buffer inventories and inflated finished goods inventory (FGI) levels to solve their flow problems. Buffers absorb the variations in demand at the high cost of additional inventories. The driving forces behind this strategy are poor responsiveness and delivery performance due to adverse setup times requiring large lot size production methods. Dramatic productivity improvements occur when close linkages among workers are coupled with management intervention to reduce buffer inventories. This is accomplished through proper setup time reduction and scheduling techniques. There are numerous benefits resulting from JIT production methods:

- Reduced lot size inventories
- Reduced buffer inventories
- Reduced finished goods inventories
- Reduced scrap
- Less direct labor wasted on rework
- Less inventory space
- Less equipment to handle inventories
- Less inventory accounting
- Less inventory control effort.

JIT is a closely linked system. Since JIT empowers the work force, the costs of administration are also lower and managers are free to deal with strategic issues. The throughput or the rate at which a company generates revenue with respect to time is also enhanced. It is possible to conceive of a situation where throughput is improved to the extent that revenue generated through sales rises more quickly than the rate at which suppliers must be paid. This phenomenon is known as *negative inventory turns.*

One of the least recognized benefits resulting from JIT production is the reduction in the adverse effects of forecast error. Forecasts are necessary to allow reduced aggregate customer and supplier lead times by building ahead.

If significant resources are expended to produce as forecasted, and the actual demand is less than forecasted, a direct loss in productivity occurs. The worst-case scenario is the situation where the lot size equals the forecast quantity. At the other extreme, if the actual demand exceeds the forecast, and the forecast lot has already been produced, the adverse consequence of an additional long setup will decrease available capacity when the difference is made up and severely impact the delivery dates of other products yet to be produced. Frequent expediting is indicative of this problem. With small-lot production, production is more linear. The forecast error is equal to the combination of lot size and the time horizon required to produce the entire mix of products. If the lot size is equal to one unit and the mix of products is produced once in one day, then the magnitude of forecast error is, at most, plus or minus one unit. This is known as the *forecast quantizing error.*

2.2 Kanban

The Kanban (pronounced *con-bon*) system of inventory control is an integral part of the JIT production system. Kanban is a Japanese word meaning "visible record" and is a tool used by manufacturers that produce in discrete units of production, such as would be found in a job shop as opposed to process industries. There are three types of Kanban systems that are of general importance. A detailed knowledge of each is required in order to select the most effective type of Kanban system for high-mix, low-volume manufacturing environments. The three types of Kanban systems are:

1. Product specific one-card Kanban system
2. Product specific two-card Kanban system
3. Generic one-card Kanban system.

For more vertically integrated manufacturers who make their components in-house, component FGI locations should be placed as close as possible to parent point-of-use (POU) stock locations. This will minimize delivery time. Additionally, the electronic Kanban signal can be replaced by a physical card and offer a more efficient and visible means by which to control the flow of material. If a product specific one-card Kanban system is used, a card is transferred to the prior process step indicating the type and quantity to be replenished. The material is subsequently replenished at the process step's interprocess stock location along with the product specific Kanban card. If a generic Kanban card is used, the prior process step determines from a schedule the type and quantity to be replenished. The generic Kanban card will be returned to the interprocess stock location along with the material delivered. It is essential to understand that at no time is material permitted to be moved

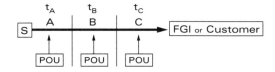

Figure 2.2.1 Time-based component or parent vector.

unless there is a Kanban card moving with it. The Kanban card is the authorization to move material and is often referred to as a *withdrawal* Kanban card.

Consider a time vector subdivided into one or more segments representing separate process steps (Fig. 2.2.1). Each process step (A, B, C) has a POU stock location for component material that is supplied directly by either internal or external vendors. The process time for each process step on the time vector is determined by the vector segment length or is explicitly indicated for each associated process step. The input to the time vector is a manufacturing schedule (S), and the output of the vector is the customer if a finished salable end product is produced; otherwise, the output is finished goods inventory (FGI). A vector will be referred to as the parent vector if its POU location is the recipient of FGI output from another time vector; otherwise, the time vector will be referred to as a component vector. We can now represent the total parent-component time-based diagram of the total value delivery chain for any particular manufacturing system by appropriately interconnecting the various component and parent time vectors.

A line connecting the FGI of one vector to the POU stock location of another vector will have a length equal to the delivery time, or the time will be explicitly indicated on the line. Of course, the value delivery chain will start from earth, the source of all our required raw materials. The overall time-based diagram will be referred to as the *concept value delivery chain* (Fig. 2.2.2). The time-based diagram is an invaluable tool to use when designing or analyzing a particular value delivery chain.

The delivery requirements are established for each component by the component vector's parent and are based on a forecast unless the parent is a build-to-order company. The demands placed on the entire concept value delivery chain by the end customer can be known by all component vectors almost immediately through electronic data interchange (EDI) means of information transmission. As each POU stock location uses material, it is replenished from the appropriate component supplier's FGI. A Kanban signal, which is a formal request for material delivery, is sent to the component supplier from the parent consumer when the parent reaches the order point. The order point is that quantity of material remaining that allows for the lead time needed to replenish the material.

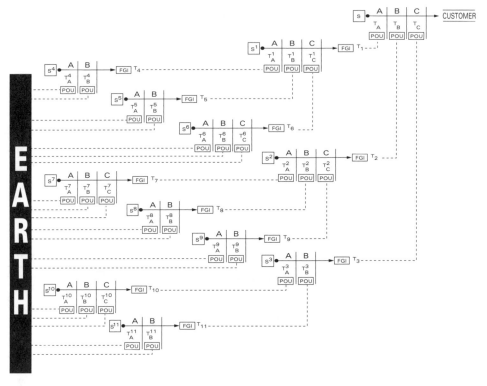

Figure 2.2.2 Concept value delivery chain.

An electronic Kanban signaling system is used between the parent and component when the physical distance between them is far enough that a physical Kanban card is impractical to transfer. The component supplier may receive a product-specific electronic Kanban signal or a physical Kanban card indicating the type and quantity to be delivered, or the component supplier could receive an electronic or physical generic Kanban signal, if the information about what to produce is derived from a schedule. The production schedule is supplied to the component supplier by its parent. The generic Kanban signal is only a signal requesting delivery and does not indicate the type or quantity to be delivered. Payment to the component supplier can also be performed electronically. Kanban systems require a good working relationship with the vendor to be effective. To facilitate this, the parent should not use multiple component vendors for a particular part number (i.e., should use a single source). Multiple component vendors would only complicate the logistics and add unnecessary cost. JIT scheduling reduces the amount of FGI that a component supplier holds due to the reduction in lot sizes. The iterative nature of JIT schedules will require the frequency of deliveries to increase.

Large batch orders that fluctuate based on demand variability would inflate the component supplier's FGI requirement and result in cost increases for the parent. Either the parent or its component supplier or both would have to maintain inflated safety stock levels to absorb the fluctuations in demand.

The parent-component Kanban system of inventory control is particularly important when we consider the adverse consequences of a machine shutdown at the parent. A Kanban signal would not be sent to the component supplier, thus preventing the unnecessary delivery of material that would only be stockpiled at the parent POU stock location. Component deliveries only occur in a Kanban system of inventory control if they are signaled for. The Kanban system of inventory control is often referred to as the *pull* system of inventory control. If the parent subsequently decides to work overtime to catch up on lost production, Kanban signals will be sent more frequently to the component supplier until the parent is back on schedule. This potential problem is the driving force behind having an excellent machine preventive maintenance program. It is now easy to understand how Kanban can effectively harness and control the variability throughout the parent-component value delivery chain, and why companies that are using JIT want their suppliers to adhere to the fundamentals of JIT production techniques.

There are clearly defined circumstances under which the choice is made as to whether a generic or product specific Kanban system should be used:

- Product specific Kanban is most appropriately used in low-mix, high-volume manufacturing environments. Low-mix, high-volume manufacturing environments facilitate the repetitive movement of material necessary to avoid carrying inventories for any significant period of time. Component vendors will continuously maintain FGI to support the volume of products produced.
- Generic Kanban is most appropriately used in high-mix, low-volume manufacturing environments. For high-mix, low-volume manufacturing environments, component inventories will not be signaled for repetitively. Some material may only be required periodically (e.g., five per month). Component vendors will not continuously maintain inventories and will typically build component inventories to order. In this case, a schedule is necessary to ensure that POU stock locations contain only those component inventories that will be used in the short term.

Now it is important to understand how Kanban is used to control the flow of inventories during the production process itself.

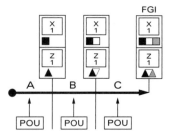

Figure 2.2.3

2.2.1 Product Specific One-Card Kanban System

Consider Fig. 2.2.3. Using a time-based component vector, let us assume that the manufacturing environment is repetitive (i.e., low-mix, high-volume) and that two products, X and Z, are going to be produced. One of each product will be stored between each of the adjacent process steps and in FGI. The time-based component vector is segmented into three process steps; A, B, and C, and the processing times for all process steps are equal. Each process step will have one worker.

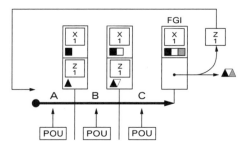

Figure 2.2.4

For the case of a product-specific one-card Kanban system, a single product specific Kanban card indicating the model number and replenishment quantity is located with each of the products throughout the system. The flow

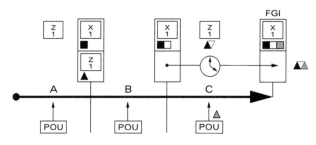

Figure 2.2.5

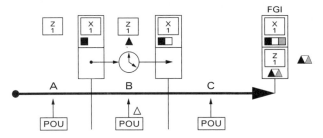

Figure 2.2.6

of inventories and product-specific Kanban cards occurs in the following sequence.

Consider Fig. 2.2.4. When the Z product is removed from FGI to be delivered to the parent, the Z Kanban card is sent to process step A. The missing product Z at FGI is a signal for process step C to produce the missing product Z.

Consider Fig. 2.2.5. Process step C will remove a partially completed product Z from the stock location at process step B along with the product Z Kanban card. The missing product Z at the process step B stock location is a signal for process step B to produce the missing Z product. After process C has completed its process, the Z product along with the Z Kanban card will be placed in FGI.

Consider Fig. 2.2.6. Process step B will remove a partially completed product Z from the stock location at process step A along with the product Z Kanban card. The missing product Z at the process step A stock location is a signal for process step A to produce the missing Z product. After process B has completed its process, the Z product along with the Z Kanban card will be placed in the process B stock location.

Consider Fig. 2.2.7. Process step A will perform the process required to create product Z from its POU stock location. After process A has completed its process, the Z product along with the Z Kanban card supplied to it from FGI will be placed in the process step A stock location.

At this point all of the produce Kanban signals are completed and the

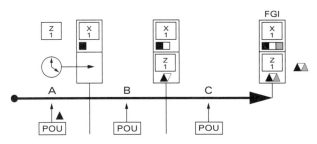

Figure 2.2.7

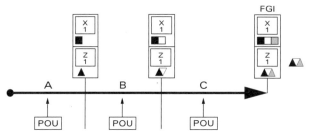

Figure 2.2.8

production system is idle (Fig. 2.2.8). Note that the Kanban cards recirculate through the production system. The only events that would retire a Kanban card from use would be if it were lost or wear and tear from constant usage required that it be replaced. For product-specific one-card Kanban systems, interprocess inventories must be maintained.

2.2.2 Product Specific Two-Card Kanban System

If the mix of products within interprocess inventories and FGI increases to the point that determining which products to replenish becomes untenable, a product-specific two-card Kanban system should be implemented. Consider the case where six products, M, V, W, X, Y, and Z, are to be produced using a one-card product-specific Kanban system. If the following sequence of products is removed from FGI, M, V, X, and Z, a worker at process step C may not be able to readily determine if another worker or machine is replenishing the particular product in question. The worker will have to ask the other workers in the work cell or inspect the feeding interprocess stock location to determine if a replenishment is occurring. This effort is time consuming and inefficient. To alleviate this problem, the product-specific two-card Kanban system was created.

Consider Fig. 2.2.9. Two Kanban cards are now attached to each product throughout the production system. One card is called the *withdrawal* Kanban

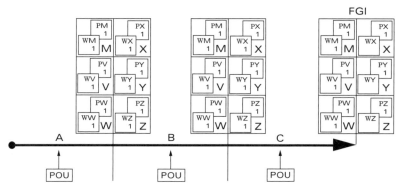

Figure 2.2.9

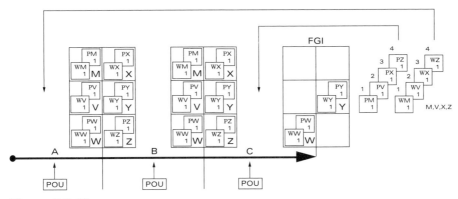

Figure 2.2.10

card, which authorizes the physical transfer of material, and the other is called the *produce* Kanban card, which authorizes the production process to begin operations. Each process step will have one worker. The flow of inventories, product-specific withdrawal, and produce Kanban cards occurs in the following sequence.

Consider Fig. 2.2.10. When the M, V, X, and Z products are removed from FGI to be delivered to the parent, the M, V, X, and Z withdrawal Kanban cards are sent to a predetermined location at process step A and the M, V, X, and Z produce Kanban cards are placed at a predetermined location at process step C. The order that the M, V, X and Z produce Kanban cards are placed at process step C will determine the order in which the products they are associated with will be replenished. The order of the M, V, X, and Z withdrawal Kanban cards sent to process step A must match the order of the M, V, X, and Z produce Kanban cards sent to process step C. The order of the withdrawal and produce Kanban cards represents the *production schedule* for replenishment.

Consider Fig. 2.2.11. Product M will be arbitrarily produced first due to simultaneous product withdrawals from FGI. Process step C will remove a partially completed product M from the stock location at process step B. The

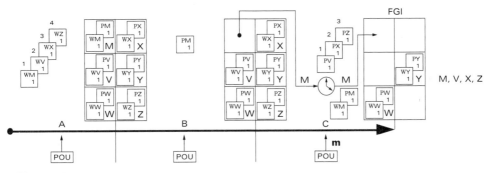

Figure 2.2.11

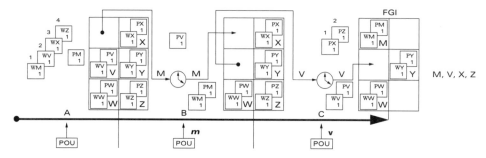

Figure 2.2.12

produce Kanban card obtained from the stock location at process step B will be sent to a predetermined location at process B while the withdrawal Kanban card will be joined with the product M produce Kanban card obtained from FGI. Process step C will complete its process on the partially completed product M obtained from the process B stock location and place the completed product M along with its withdrawal and produce Kanban cards into the appropriate FGI stock location.

Consider Fig. 2.2.12. Process step B will remove a partially completed product M from the stock location at process step A. The produce Kanban card obtained from the stock location at process step A will be sent to a predetermined location at process step A, while the withdrawal Kanban card will be joined with the product M produce Kanban card obtained from process step C. Process step B will complete its process on the partially completed product M obtained from the process step A stock location and place the completed unit along with its withdrawal and produce Kanban cards into the appropriate process step B stock location. Simultaneously, in time with respect to process step B, process step C will remove a partially completed product V from the stock location at process step B. The produce Kanban card obtained from the stock location at process step B will be sent to a predetermined location at process step B, while the withdrawal Kanban card will

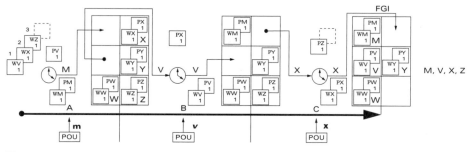

Figure 2.2.13

be joined with the product V produce Kanban card obtained from FGI. Process step C will complete its process on the partially completed product V obtained from the process step B stock location and place the completed product V along with its withdrawal and produce Kanban cards into the appropriate FGI location.

Consider Fig. 2.2.13. Process step A will join the product M produce Kanban card obtained from process step B with the product M withdrawal Kanban card obtained from FGI and perform the process required on product M. The completed product M and its withdrawal and produce Kanban cards are placed in the appropriate process step A stock location. Simultaneously with process step A, process step B will remove a partially completed product V from the stock location at process step A. The produce Kanban card obtained from the stock location at process step A will be sent to a predetermined location at process step A at the same time that the withdrawal Kanban card will be joined with the product V produce Kanban card obtained from process step C. Process step B will complete its process on the partially completed product V obtained from the process step A stock location and place the completed product V along with its withdrawal and produce Kanban cards into the appropriate process step B stock location. Also occurring simultaneously with process steps A and B, process step C will remove a partially completed product X from the stock location at process step B. The produce Kanban card obtained from the stock location at process step B will be sent to a predetermined location at process step B, while the withdrawal Kanban card will be joined with the product X produce Kanban card obtained from FGI. Process step C will complete its process on the partially completed product X obtained from the process step B stock location and place the completed product X along with its withdrawal and produce Kanban cards into the appropriate FGI location.

If product M were now removed from FGI, its withdrawal Kanban card would be sent to process step A, and its produce Kanban card would be sent to process step C. We want to ensure that the product M withdrawal Kanban card would be placed after the product Z withdrawal Kanban card at process step A, and the product M produce Kanban card would be placed after the product Z produce Kanban card at process step C. Failure to do this would result in an adversely long replenishment time for product Z. The production process would become Last In, First Out (LIFO) rather than First In, First Out (FIFO) priority control.

While high quality POU component inventories delivered Just-In-Time support the uninterrupted flow of material through the manufacturing process, many manufacturing managers have the mistaken belief that the production process should respond Just-In-Time to arbitrary external demands placed upon the production system. The ability to respond to external demand is

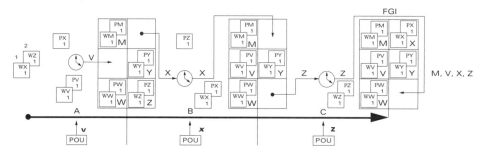

Figure 2.2.14

strictly a scheduling issue and not a Just-In-Time issue. JIT will optimize the utilization of *available* capacity through the elimination of waste but JIT will not create more capacity than is available. If demand grows to the point that it exceeds available capacity, investment is required to increase capacity.

Consider Fig. 2.2.14. Process step A will join the product V produce Kanban card obtained from process step B with the product V withdrawal Kanban card obtained from FGI and perform the process required on product V. The completed product V and its withdrawal and produce Kanban cards are placed in the appropriate process step A stock location. Simultaneously with process step A, process step B will remove a partially completed product X from the stock location at process step A. The produce Kanban card obtained from the stock location at process step A will be sent to a predetermined location at process step A; meanwhile the withdrawal Kanban card will be joined with the product X produce Kanban card obtained from process step C. Process step B will complete its process on the partially completed product X obtained from the process step A stock location and place the completed product X along with its withdrawal and produce Kanban cards into the appropriate process step B stock location. Also occurring simultaneously with process steps A and B, process step C will remove a partially

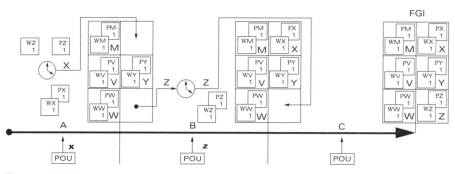

Figure 2.2.15

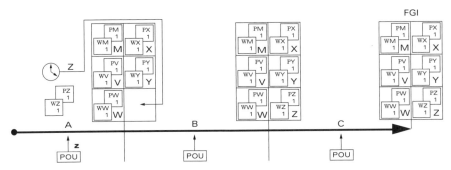

Figure 2.2.16

completed product Z from the stock location at process step B. The produce Kanban card obtained from the stock location at process step B will be sent to a predetermined location at process step B while the withdrawal Kanban card will be joined with the product Z produce Kanban card obtained from FGI. Process step C will complete its process on the partially completed product Z obtained from the process step B stock location and place the completed product Z along with its withdrawal and produce Kanban cards into the appropriate FGI location.

At this point, the FGI stock location is completely replenished (Fig. 2.2.15). Process step A will join the product X produce Kanban card obtained from process step B with the product X withdrawal Kanban card obtained from FGI and perform the process required on product X. The completed product X and its withdrawal and produce Kanban cards are placed in the appropriate process step A stock location. Simultaneously, in time with respect to process step A, process step B will remove a partially completed product Z from the stock location at process step A. The produce Kanban card obtained from the stock location at process step A will be sent to a predetermined location at process step A while the withdrawal Kanban card will be joined with the product Z produce Kanban card obtained from process step C. Process step B will complete its process on the partially completed product Z obtained from the process step A stock location and place the completed product Z along with its withdrawal and produce Kanban cards into the appropriate process step B stock location.

At this point, the process B stock location is completely replenished (Fig. 2.2.16). Process step A will join the product Z produce Kanban card obtained from process step B with the product Z withdrawal Kanban card obtained from FGI and perform the process required on product Z. The completed product Z and its withdrawal and produce Kanban cards are placed in the appropriate process step A stock location. At this point all of the produce Kanban signals are completed and the production system is idle.

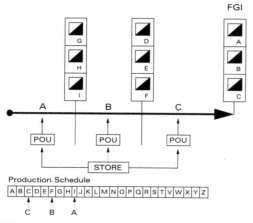

Figure 2.2.17

It should now be clear that product-specific Kanban is applicable to repetitive manufacturing environments and not low-volume, high-mix manufacturing environments due to the requirement for fixed and continuously maintained interprocess inventories.

2.2.3 Generic One-Card Kanban System

Generic Kanban is the most effective means of inventory control for a high-mix, low-volume manufacturing environment. In such an environment, the contents of all POU, interprocess, and FGI stock locations are based on scheduled order mix and will change over time. Vendors will no longer deliver material directly to POU locations. Inventory will not be used repetitively and fixed POU stock locations no longer exist—a store requirement. A store serves the purpose of receiving vendor materials, stocking them, and ensuring that the right components are delivered to the right POU locations at the

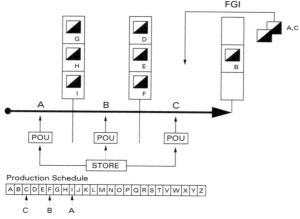

Figure 2.2.18

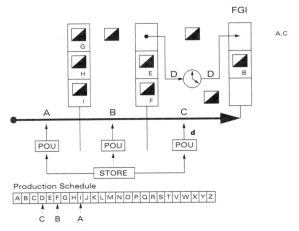

Figure 2.2.19

right time, based on the production schedule. The flow of inventories and generic Kanban cards occurs in the following sequence:

Consider Fig. 2.2.17. The initial conditions are established for the high-mix schedule. All process steps, POU stock locations, and stores operate from a common schedule derived from actual or forecast orders and each process step has one worker.

Consider Fig. 2.2.18. A generic Kanban card is both a production authorization and withdrawal authorization card. When the A and C products are removed from FGI to be delivered to the parent, the A and C generic Kanban cards are sent to a predetermined location at process step C. The number of generic Kanban cards represents the number of products that process C will produce.

Consider Fig. 2.2.19. Based on the production schedule, process step C will

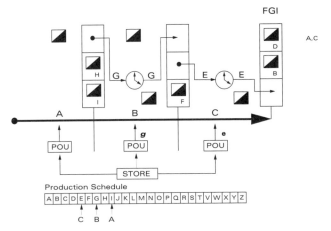

Figure 2.2.20

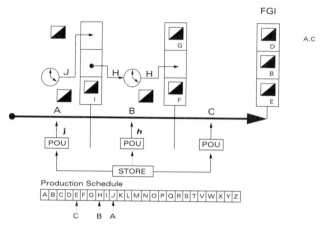

Figure 2.2.21

increment its schedule and subsequently remove a partially completed product D from the process step B stocking location. The generic Kanban card obtained from the process step B stock location will be sent to a predetermined location at process step B while the generic Kanban card obtained from FGI will be joined with the partially completed product D. Process step C will complete its process on the partially completed product D obtained from the process step B stock location and place the completed unit along with its generic Kanban card into any available FGI stock location.

Consider Fig. 2.2.20. Based on the production schedule, process step B will increment its schedule and subsequently remove a partially completed product G from the stock location at process step A. The generic Kanban card obtained from the process step A stock location will be sent to a predetermined location at process step A while the generic Kanban card obtained from process step C will be joined with the partially completed product G.

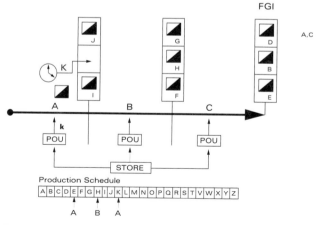

Figure 2.2.22

Process step B will complete its process on the partially completed product G obtained from the process step A stock location and place the completed unit along with its generic Kanban card into any available process step B stock location. Based on the production schedule and simultaneously with process step B, process step C will increment its schedule and subsequently remove a partially completed E product from the stock location at process step B. The generic Kanban card obtained from the process step B stock location will be sent to a predetermined location at process step B, while the generic Kanban card obtained from FGI will be joined with the partially completed product E. Process step C will complete its process on the partially completed unit obtained from the process step B stock location and place the completed unit along with its generic Kanban card into any available FGI stock location.

Consider Fig. 2.2.21. At this point the FGI stock location is completely replenished. Based on the production schedule, process step A will increment its schedule and subsequently perform the process required on product J. The generic Kanban card obtained from process step B will be joined with the completed unit and placed into any available process step A stock location. Based on the production schedule and simultaneously with process step A, process step B will increment its schedule and subsequently remove a partially completed H product from the stock location at process step A. The generic Kanban card obtained from the process step A stock location will be sent to a predetermined location at process step A, while the generic Kanban card obtained from process step C will be joined with the partially completed product H. Process step B will complete its process on the partially completed unit obtained from the process step A stock location and place the completed unit along with its generic Kanban card into any available process step B stock location. At this point the process step B stock location is completely replenished.

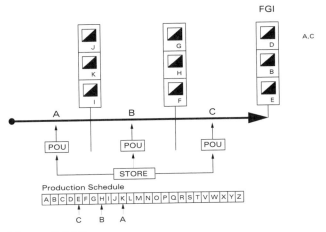

Figure 2.2.23

Consider Fig. 2.2.22. Based on the production schedule, process step A will increment its schedule and subsequently perform the process required on product K. The generic Kanban card obtained from process step B will be joined with the completed unit and placed into any available process step A stock location.

At this point all of the generic Kanban signals are completed and the production system is idle (Fig. 2.2.23). Although a generic Kanban card can be used, it is also possible to use a light at each process step to indicate when production should begin. The light at each process step would be controlled by the subsequent process step or FGI. The light would illuminate to indicate that production should stop, otherwise production should continue. The Japanese call such lights *andon*. Another alternative would be to simply let an empty stock location indicate that production should occur. There are many creative signaling methods used today to facilitate effective Kanban inventory control technique. Kanban is a production activity control and material delivery issue, not an end-customer order administration issue. Kanban is a tool that is used for controlling the delivery of internal or external component inventories as well as for controlling interprocess inventories and FGI. The end customer never sees Kanban. Every manufacturer in the world receives orders from customers. If the placement of external orders at a production facility were what constituted a Kanban system, every manufacturer in the world could claim to have one.

2.3 Theory of Constraints

Theory of constraints is a manufacturing model developed by Eliyahu M. Goldratt[48,49] which recognizes that constraints will determine the overall performance of a manufacturing system. A constraint is defined as anything that limits performance relative to the goals of making money now and in the future. The total number of constraints in a manufacturing environment is

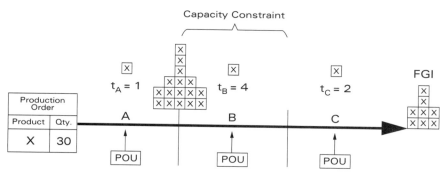

Figure 2.3.1 Theory of constraints.

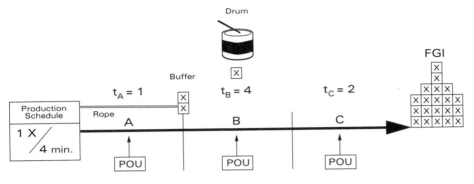

Figure 2.3.2 Theory of constraints.

small. By improving and mitigating constraints, the overall system is improved.

Consider a manufacturing time vector subdivided into three process steps, A, B, and C (Fig. 2.3.1). A single product, X, is manufactured and the total process time is 1, 4, and 2 time units for process steps A, B, and C respectively. The setup time for each process step is one time unit. If production orders for product X were launched unabated into this factory, a queue would appear at the input of process step B. This is referred to as *blocking*.

Theory of constraints refers to process step B as a capacity constraint. Process step B will exhibit reduced throughput as a result of a machine breakdown at process step A. Consider Fig. 2.3.2. To protect the constraint from periods of starvation, a time buffer is placed at the input of the constraint. The time buffer contains the quantity of X units that translate into a particular amount of process time at the constraint. The rate that the constraint processes units (analogized as a *drum* beat) contained in the *buffer* is synchronized with the rate of replenishment. The quantity required to replenish the buffer is forward scheduled into the gating process step, A. The linkage between the time buffer and the gating operation is analogized as a *rope*. The time buffer is managed based on the average demand placed on the time buffer. If the constraint empties the buffer before it is replenished, the buffer is expanded. If excess inventory is continuously present in the buffer, the size of the buffer is reduced. The output of this production system is limited absolutely by the constraint. If scheduled maintenance or a machine breakdown occurred at the constraint, throughput would be adversely affected. The throughput of the production system is protected from this occurrence by placing a time buffer at the output of the constraint. Improvements at the constraint such as reduced setup time, improved preventive maintenance, more efficient equipment, etc., will result in a consequent improvement in throughput and the inventory level for the buffers will be reduced proportionately. Theory of constraints is often referred to as the *drum, buffer, rope,* production system.

If three products, X, Y, and Z, were produced by this production system, the contents of the time buffer would change with respect to time based on a production schedule. The contents of the time buffer would contain the mix of products representing a fixed scheduled period of time.

In the course of managing the actual versus the planned contents of the time buffer, deviations will drive the focus of continuous improvement efforts. The time buffers contain most of the inventory and are used to protect the factory against disruptions. Inventory at any other location would be a waste. If disruptions occur, the buffer(s) will always be smaller than planned. To maximize system throughput, every effort is made to utilize a constraint at maximum capacity.

Theory of constraints will prioritize the schedule of products by giving higher profit margin products preference over lower margin products. Lower margin products will be delivered last. This is generally the case unless external market factors dictate otherwise. For instance, a lower margin product delivered sooner may leverage sales of higher margin products in the future. Flexibility in scheduling is critical in this regard.

High-mix manufacturing environments are characterized by different products using the same manufacturing resources. For dissimilar products, the processing times at each process step will often differ. One or more capacity constraints may exist and *shift* based on the mix of products being produced over a particular time horizon. Theory of constraints does not recognize this fact and is only effective to use when the number of constraints is fixed and robust to the incoming order mix. For high-mix manufacturing environments, the likelihood of constraints remaining stationary is significantly reduced.

Consider a time vector subdivided into three process steps: A, B, and C. Three products, X, Y, and Z, will be produced such that the total time, expressed in time units, to process each product at each process step is as follows (see Table 2.3.1):

Table 2.3.1 Process time

	X	Y	Z
Process A	4	1	2
Process B	2	8	2
Process C	1	1	6

Table 2.3.2 Order mix 1

	Order Qty.	% Total
Product X	200	20
Product Y	500	50
Product Z	300	30
Total	1,000	100

200(X)	500(Y)	300(Z)

Schedule 1

If the orders for all products were as indicated in Table 2.3.2, and we were to produce all products using schedule 1, we would exhibit extremely poor customer responsiveness, and the process constraint would shift based on the total processing times (see Table 2.3.3):

Table 2.3.3 Total processing time

	Process A	Process B	Process C
Product X	800*	400	200
Product Y	500	4,000*	500
Product Z	600	600	1,800*

* denotes a capacity constraint.

The capacity constraint would shift from process A to process B to process C. Since process B represents the greatest total processing time, it would manifest itself as the bottleneck process. If the incoming order mix were to shift due to a major marketing campaign to increase the sales of product Z, the following incoming order mix might be observed (see Table 2.3.4):

Table 2.3.4 Order mix 2

	Order Qty.	% Total
Product X	200	20
Product Y	200	20
Product Z	600	60
Total	1,000	100

If we were to produce all products using the schedule:

200(X)	200(Y)	600(Z)

Schedule 2

we would again exhibit extremely poor customer responsiveness, and the process constraint would shift based on the total processing times (see Table 2.3.5):

Table 2.3.5

	Process A	Process B	Process C
Product X	800*	400	200
Product Y	200	1600*	200
Product Z	1,200	1,200	3,600*

* denotes a capacity constraint.

The constraint would again shift from Process A to Process B to Process C; however, Process C would now represent the greatest total processing time and manifest itself as the bottleneck process.

For high-mix, low-volume manufacturing environments, it is not economical to product-focus production due to the expense of machines and additional factory space required. Excess capacity would exist due to the low volumes of products produced. For this reason, high-mix, low-volume manufacturing environments should be process-focused using flexible equipment with setup times that will confer a competitive advantage. Many manufacturers attempt to enhance the flexibility of machines ill-equipped to obtain the setup time reductions necessary to compete effectively. While setup time reductions may be achieved, the likelihood is remote that they will be equivalent or superior to those of a competitor who is using equipment specifically designed for setup time flexibility and is a risk that must be avoided.

As the diversity of products produced increases along with the volatility of incoming order quantities, the probability of moving constraint and bottleneck operations is increased. For high-mix, low-volume manufacturing environments, mix can be as high as 600 different products or greater, and incoming order quantities may range anywhere from 0-1,000 or greater for individual products within the mix over a particular time horizon. Manufacturers that produce in batches will experience the problem of moving constraints and bottlenecks. This phenomenon may vary with a company's financial performance. This is why poor financial performance is often blamed on an unfavorable order mix. Unfavorable order mix is a symptom rather than the cause of poor

Table 2.3.6

1	2	3	4	5	6	7	8	9	10	% Demand
Y		Y		Y		Y		Y		50 Model Y
	Z				Z				Z	30 Model Z
			X				X			20 Model X

Y	Z	Y	X	Y	Z	Y	X	Y	Z	Schedule 1a
B	C	B	A	B	C	B	A	B	C	Constraint
8	6	8	4	8	6	8	4	8	6	Constraint time

financial performance. The management-controlled production system is flawed. Through proper scheduling techniques, the manufacturing production system can be made relatively robust to incoming order mix.

With significantly reduced setup times, moving constraints and bottlenecks become increasingly problematic if we attempt to increase our responsiveness to the customer by creating a small lot size yet an improper schedule. Consider schedule 1a in Table 2.3.6, which is derived from order mix 1. The constraint is continually moving. For this example, expensive buffer inventory would be required at every process step based on the aforementioned theory of constraints approach. Each process is being alternately blocked and starved. Another approach to the scheduling problem would be to consider the total processing time for each product and calculate the least common denominator in time for each product across the entire mix of products (see Table 2.3.7).

Table 2.3.7

	Total Processing Time
Product X	7
Product Y	10
Product Z	10

$$1/7 + 1/10 + 1/10 = 10/70 + 7/70 + 7/70$$
$$X + Y + Z = X + Y + Z$$

Schedule 1b → X:X:X:X:X:X:X:X:X:X Y:Y:Y:Y:Y:Y:Y Z:Z:Z:Z:Z:Z:Z

Constraint → A:A:A:A:A:A:A:A:A:A B:B:B:B:B:B:B C:C:C:C:C:C:C

Constraint Time → _____40_____ |_____56____ |_____42_____

Total Time → _____70_____ |_____70____ |_____70_____

Although each product spends the same total amount of time in the production system, production is not synchronized with the constraints. The constraints will still be alternately blocked, which inflates inventory, or starved, which results in lost throughput. For schedule 1b, constraint C will be starved for 14 time units and constraint B will be blocked for 16 time units. The nonconstraint process steps should only be activated for the purpose of satisfying the constraint process steps. The throughput of the overall system is controlled by the constraint process steps; therefore, we proceed as follows to obtain synchronization across all constraint process steps (see Tables 2.3.8 and 2.3.9):

Table 2.3.8

	Constraint Process Time
Product X	4
Product Y	8
Product Z	6

$$1/4 + 1/8 + 1/6 \; = \; 6/24 + 3/24 + 4/24$$
$$X + Y + Z \; = \; X + Y + Z$$

Table 2.3.9

X	X	X	X	X	X	Y	Y	Y	Z	Z	Z	Z	Schedule 1c
A	A	A	A	A	A	B	B	B	C	C	C	C	Constraint
24						24			24				Constraint time

As a result of developing a schedule that synchronizes production across all of the constraints, the adverse effects of blocking and starvation of the constraints have been eliminated.

The mathematical basis for schedule 1c is easily formulated. The total time to produce products X, Y, and Z can be mathematically expressed as:

$$T_t = T_c + \sum_{i=1}^{n} T_{\bar{c}_i} \qquad\qquad (2.3.1)$$

Where: T_t = Total process time
        $T_{\bar{c}}$ = Nonconstraint time
        T_c = Constraint time
        n = Total number of nonconstraint operations.

The total time is equal to the constraint time plus the summation of all nonconstraint times. One unit will be produced if one worker or machine is used at each process step. This can be expressed as:

$$\frac{Units}{Time} = \frac{1}{T_c + \sum\limits_{i=1}^{n} T_{\bar{c}_i}} \qquad (2.3.2)$$

The nonconstraints will not affect the cycle time and are thus equal to zero. The relationship can now be expressed as:

$$\frac{Units}{Time} = \lim\limits_{\sum T_{\bar{c}} \to 0} \left(\frac{1}{T_c + \sum\limits_{i=1}^{n} T_{\bar{c}_i}} \right) = \frac{1}{T_c} \qquad (2.3.3)$$

The production schedule is based on the summation of the reciprocals of the individual process constraint times for each product produced:

$$Schedule = \sum\limits_{j=1}^{k} \frac{1}{T_{c_j}} = \frac{1}{T_{c_1}} + \frac{1}{T_{c_2}} + \ldots + \frac{1}{T_{c_k}} \qquad (2.3.4)$$

Where k is equal to the total number of constraints.

The system's least common denominator *(LCD)* is now determined so that it is not less than the greatest constraint time:

$$Schedule = \sum\limits_{j=1}^{k} \frac{N_j}{LCD} = \frac{N_1}{LCD} + \frac{N_2}{LCD} + \ldots + \frac{N_k}{LCD} \qquad (2.3.5)$$

Therefore:

$$Schedule: N_1, N_2, \ldots, N_k$$

The schedule will be iterated until the ordered quantities are exhausted. When the quantity of a particular product associated with a particular constraint is exhausted and the remaining products associated with other constraint operations are not, the LCD is recalculated and the schedule is updated. This may change the lot sizes of the remaining products. This procedure is known as *dynamic lot sizing*. In the case where the quantity of a particular product type may not be sufficient to satisfy the quantity required by the schedule for a particular constraint, the remnants will be produced and not disrupt the overall schedule. This situation may occur when the quantity of a particular product is nearing completion. The result is a schedule that is precisely equal to actual order quantities.

It is important to understand that schedule 1 and schedule 2 may incorrectly lead management to believe that an increase in orders will always require additional inventories. It should be clear that the iteration of schedule 1c will result in a relatively level inventory requirement that is significantly lower than either schedule 1 or schedule 2, despite the volatility of incoming orders. Believing that inventory levels are determined by incoming orders is dangerous if thought of as normal or acceptable. Excess capacity is usually required to successfully respond to the dynamics of incoming orders when operations are scheduled in the manner of schedule 1 and schedule 2. Schedule 1c is scheduled in a manner that utilizes the constraint resources at maximum capacity, only allowing excess capacity for the purpose of responding to disruptions (e.g., scheduled maintenance, absenteeism, etc.). Schedule 1c smoothes and levels the load while schedules 1 and 2 do not. The result is a highly responsive, high throughput, and low inventory production environment.

Reduced setup times coupled with proper scheduling will decouple manufacturing from incoming order volatility without a capacity buffer (i.e., excess capacity). Manufacturing is robust to incoming order volatility when this occurs. It is important to note that schedule 1c repeatedly negotiates the entire mix of products thereby profit margins will be relatively balanced. This is not the case for schedules 1 and 2. Consider the case where the profit margins for products X, Y, and Z are $1,000, $3,000, and $2,000 respectively. The first thirteen products produced by schedules 1 and 2 generate $13,000 in profits while schedule 1c generates $23,000 in profits. The smooth return on manufacturing investment versus time for schedule 1c is superior to the lumpy return on manufacturing investment for schedules 1 and 2.

Many manufacturing managers believe that unfavorable order mix is the cause of poor financial results when, in fact, the manufacturing system they control is the root cause. This is another classic example of confusing the symptom with the cause. Intuition is no substitute for a thorough understanding of manufacturing fundamentals.

2.4 Available Capacity

In order to effectively achieve customer acknowledge dates (i.e., delivery dates), capacity planning is a must. Production plans derived from schedules are of little use when setting acknowledge dates if acknowledge dates are determined based on overloaded production operations due to poor or nonexistent capacity planning. While scheduling is used to smooth demand, load leveling is the balancing of load with capacity. A level load is the result of proper capacity planning. Capacity is defined as the rate at which work is output by the production system. Available capacity is most appropriately defined as follows:

Available capacity = Time available × *efficiency* × *availability* × *activation*   (2.4.1)

Utilization = Availability × *activation*   (2.4.2)

Efficiency = Standard units of time produced / time available   (2.4.3)

Availability = 1- (time down / time available)   (2.4.4)

Activation = 1 - [idle time / (time available - time down)]   (2.4.5)

Based on these calculations, acknowledge dates can be determined with confidence. The manufacturing plan will be achievable if the load is balanced with available capacity. From a JIT management perspective, availability and activation should be used as production process performance measures. The ideal availability will be equal to one. Setup time and scheduled or unscheduled maintenance are major contributors to machine availability being suboptimal. Setup time and unscheduled machine downtime should be aggressively reduced. Activation will not as a general rule be equal to one. When the processing times for operations process steps are not equal, a capacity constraint will exist. A capacity constraint is defined as the process step where a particular product or subassembly requires the greatest amount of time for performance of value-adding production operations. Value-adding time is often referred to as work content time. The full activation of a nonconstraint is counter-productive. The optimal activation of a nonconstraint is determined by the capacity constraint. Only the capacity constraint operation should have an activation equal to one. In the special case where the work content times for all process steps are equal, an activation of one is acceptable for all process steps. It is important to note that efficiency is very sensitive to the experience and training of the workers present. Demonstrated (i.e., historical) capacity calculations should not be used. Demonstrated capacity has a tendency to hide rather than expose production problems. This will hamper continuous improvement efforts.

2.5 Production System Types

The positioning of a production system is performed in two dimensions. A production system is positioned by system type and finished goods inventory policy (Fig. 2.5.1).

Positioning a production system is of strategic importance due to the critical linkage it provides to the marketplace. A key consideration when positioning a manufacturing production system is the product life cycle. Product life cycles drive marketing sales strategies and pricing strategies that, in turn, drive strategic manufacturing decisions.

The types of production systems that can be chosen are product-focused and process-focused. The type of production system selected for high-mix manufacturing environments must be flexible to respond to intermittent demand. Process-focused systems apply to a broad spectrum of product mix where changeover flexibility is needed. The capacity needs for high-mix, low-volume manufacturing environments do not justify facilities that are dedicated to a single product or product family. Advanced technologies are integral to the process-focused positioning strategy. The technologies adopted must provide new capabilities or enhance process capabilities to obtain competitive advantage. The optimal technology for high-mix manufacturing is multipurpose machinery that can be quickly changed over, is highly reliable, and easy to maintain. The requirement for high quality will require a very tight machine tolerance capability.

In contrast to process-focused is product-focused. Product-focused production systems are based on continuous or repetitive demand produced on a dedicated production line. Volume-focusing is a specialization of product-focusing that carries the high risk of inflexibility. Changes in product mix or product life cycle changes will leave the manufacturing facility vulnerable to more flexible competitors. There is little likelihood that a product produced

	Build-to-order	**Build-to-stock**
Product-focused	Low-mix, High-volume	Low-mix, High-volume
Process-focused	High-mix, Low-volume	High-mix, Low-volume

Figure 2.5.1 Manufacturing positioning.

by a volume-focused production line will remain static throughout its life cycle. The product life cycle at introduction is usually based on innovation and starts with relatively low volumes. The growth phase is characterized by increasing volumes where variety, quality, responsiveness, and cost are competitive differentiators. At the product maturity and decline phase, order volumes decrease and competition is usually driven by availability and cost. For high-mix, low-volume manufacturing environments that are volume or product-focused, expensive repositioning will be required to align production with changes in the product life cycle.

The finished goods inventory policies of build-to-stock or build-to-order manufacturing systems are the other key considerations for positioning a manufacturing environment. Build-to-stock is typically selected to offer improved customer responsiveness. Build-to-stock is expensive and finished goods inventory levels must be forecast to maintain the level of service required at the minimum investment possible. The finished goods inventory (FGI) position selected is not dependent upon the physical system that is either process- or product-focused. For build-to-stock, the finished goods inventory level should be proportional to work-in-process (WIP) inventory. WIP inventory and production lead time are the same thing. With significant reductions in WIP inventory, a build-to-stock company can almost seamlessly be converted to a build-to-order company. Although WIP inventory levels do not directly determine FGI levels, manufacturing management must set FGI levels based on WIP levels. WIP levels will thus be synchronized in time with FGI replenishment time requirements. The FGI level will be somewhat more than WIP levels in order to buffer incoming order volatility. Where immediate delivery is not required, build-to-order is the preferred position in either process- or product-focused manufacturing environments due to savings from reduced inventory levels that occur relative to a build-to-stock position.

The management systems for planning and controlling functions are different for each positioning decision. For the hybrid approach of product- and process-focusing, a factory will increase costs due to the requirement of increased overhead to facilitate separate planning functions. Ineffective positioning will result in an ineffective manufacturing strategy. The strategic importance of positioning requires that marketing and manufacturing management work together to form an integrated overall strategy that will successfully satisfy the market requirements for the entire product line manufactured. For high-mix, low-volume manufacturing environments, process-focused, build-to-order is the most cost-effective positioning decision.

When making manufacturing position choices, core competencies should not be changed dramatically. Restaffing and long learning curves are usually associated with major shifts in core competencies. This could occur if, for example, a low-mix, high-volume manufacturer attempted to shift from a product-focused repetitive demand factory to a high-mix, low-volume intermittent demand factory.

2.6 Manufacturing Resource Planning (MRP II)

MRP II is a computer-based system that integrates all facets of a manufacturing company and is essential for effectively managing a high-mix, low-volume manufacturing environment. At the executive level, MRP II integrates the business planning elements of marketing, finance, and manufacturing. MRP II also provides a means of monitoring and updating information relative to operations and purchasing functions through feedback channels. MRP II is a formal system that extends material requirements planning (MRP) by effectively planning and scheduling resources and simulating production options. Fig. 2.6.1 shows the various components of MRP II.

Successful high-mix, low-volume manufacturing organizations require effective top management planning. Top management planning begins with the mission statement. The mission statement is a document that communicates the purpose of the organization over a period of several years. The mission statement will typically define what markets and customer base are to be served and will position the organization in a way that capitalizes on the company's strengths. The mission statement is usually accompanied by a statement of guiding beliefs, values, and principles that will focus the organization on the achievement of its mission.

Business planning is a long range strategic activity performed annually that establishes the goals that will support the mission of the company. The output of the business planning process is a business strategy statement (BSS) that addresses the following issues:

- What is the basis for achieving competitive advantage (e.g., quality, cost, delivery, responsiveness, service, technology)?
- What are the financial strategies as pertains to profit, cash flow, return on investment, return on assets, and growth, based on the balance sheet and income statement?
- What are the resource requirements (e.g., technology, work force skills, plant and equipment, distribution, capital and debt financing required to obtain the assets necessary to achieve planned sales volume, departmental budgets, growth)?

The business planning process commits the company's present and future resources in a way that is responsive to the external marketplace. Defining the company's mission and business plan is a top management responsibility that

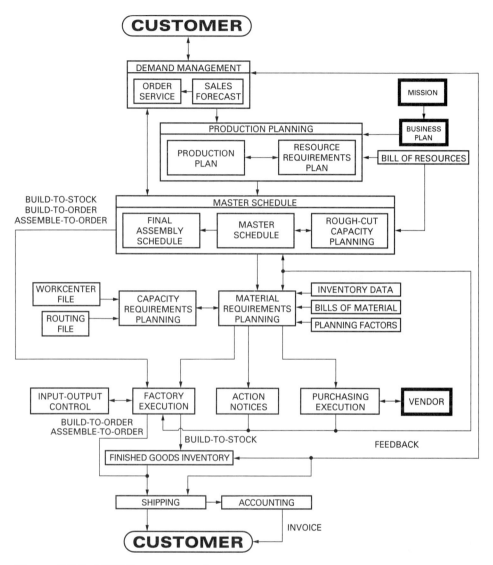

Figure 2.6.1 MRP II components.

involves top executives from finance, engineering, marketing, and manufacturing. All decisions at the business planning level comprise specific measur-
able objectives and action plans to achieve those objectives consistent with the
mission of the firm. Effective business planning requires creative and imagi-
native participants who can meet the competitive challenges of a dynamic
marketplace.

Following the business planning process is the production planning
process (Fig. 2.6.1). Production planning couples the business plan and sales

forecast with the master production schedule. Production planning is performed on a monthly basis for a one-year time horizon and reconciles aggregate monthly production levels with sales and inventory objectives. The production plan is the formal management authorization to develop a more detailed master production schedule. The production planning meeting is chaired by the general manager, and staff managers from sales, marketing, manufacturing, engineering, finance, materials, and quality should be in attendance. The production planning process measures and evaluates the company's performance in terms of sales, production, inventory, new product introductions, warranties, and monthly production defect rates in aggregate on a monthly basis. The result of the production planning process is the setting of monthly goals for production rate, inventory level, and capacity level. The information required to effectively perform the production planning process should be stratified by product families and include the following:

- Actual and planned shipments
- Customer order rates
- Order backlog
- Sales forecast
- New product introductions
- Inventory status (raw material, work-in-process, finished goods inventory)
- Actual and planned production output
- Financial constraints (e.g., capital investment or debt limitations)
- Material constraints (e.g., purchase part supply or lead time limitations)
- Warranty level
- Manufacturing defect level or turn-on rate at test.

It is important to create the minimal number of product family groupings possible to gain efficient and effective control from management. It is simply not practical for management to consider every item that the company produces. Product family groupings are analyzed based on the aggregate subset of products that are build-to-order and the aggregate subset of products that are build-to-stock. Monthly production rates are determined for build-to-stock, build-to-order, and assemble-to-order products using the following calculations:

Build-to-stock:
Production plan = Forecast + ending inventory - beginning inventory   (2.6.1)

Production rate = Production plan / number of monthly periods   (2.6.2)

Build-to-order:

Production plan = Forecast + beginning backlog - ending backlog   *(2.6.3)*

Production rate = Production plan / number of monthly periods   *(2.6.4)*

Assemble-to-order:

Production plan = Forecast + semifinished goods inventory change - backlog change   *(2.6.5)*

Production rate = Production plan / number of monthly periods   *(2.6.6)*

These calculations set the monthly production rates based on management consideration of desired semifinished goods inventory, finished goods inventory, and order backlog levels. In order to monitor and evaluate production plan performance criteria, actual versus planned levels for sales, production, inventory, backlog, and shipments must be charted. Monthly absolute deviations as well as the month-to-month cumulative deviations will greatly assist management during the production planning process. The aggregate levels for all product family groupings should compute to the business plan. The business plan operating budget represents the total business volume for the year stated in dollars. If product families are stated in units, the dollars they represent should add up to the business plan. Due to the dissimilarity of products typically found in a high-mix, low-volume manufacturing environment, product families should be expressed in dollars.

After the production plan is determined, it is critical to validate that adequate resources are available. Essentially, we want to resolve any conflicts that will prevent the production plan from balancing supply and demand. Resource requirements planning ensures that adequate production capacity is available and is based on a bill of resources (Fig. 2.6.1). The bill of resources specifies how much of a critical resource (i.e., capacity constraint) is required to produce each unit for each product family. Resource requirements should be time-phased based on the production lead time. The start date is derived by offsetting the production lead time from the completion date. Due to the level of sophistication involved, the use of time-phased bill of resource profiles for the purpose of production capacity planning should be performed by computer spreadsheet analysis. Computer analysis also facilitates efficient "what if" analyses when reconciling production capacity with the production plan. When production lead times are sufficiently short (e.g., days), time-phasing the resource requirements plan is not a necessity. It is important to understand that other constraints such as limited financial resources, labor shortages, or limited material availability can also prevent the production plan from being realized. The production plan is not a wish list. The production plan must be developed in a way that does not violate the imposed constraints.

By adjusting planned production, inventory, backlog levels, and available capacity, overloads and underloads can be effectively addressed.

Production plan time fences must be established in order to effectively manage change. When a decision is made to increase available capacity, strategies such as outsourcing, expanding the work force, or purchasing additional machinery will take time. The time fence will ensure that excessive costs are not incurred as a result of attempting changes within the time required to implement the desired change and will also prevent the overloading of the production schedule if the required change is impossible to achieve. Long lead time material delivery or limited material availability are also important considerations when determining time-fence type (e.g., capacity time fence, material time fence) as well as placement. Time fencing helps prevent management from making changes that are ineffectual or detrimental.

Production planning is vital if the mission, business plan, and master plan are to be achieved. Production planning is a formal top management process that allows management to gain and keep control of the business. The objectives, monthly meeting schedule, membership, product families, planning horizon, time fences, preparation process, and review process must be established as company policy. The benefits of production planning are numerous:

- Excellent communication
- Management accountability
- Effective change management
- Achievement of return on investment goal
- Achievement of profit goal
- Increased manufacturing productivity
- Reduced inventory levels
- Reduced operating costs
- Improved delivery performance.

Demand management is the process of managing all independent demands for a company's product line and effectively communicating these demands to the master planner and top management production planning function (Fig. 2.6.1). A key function of demand management is customer order service. Customer order service methods are driven by the types of production environments employed for the various product lines offered to the customer. There are three types of production environments that are generally encountered, and they directly affect the ability to provide the customer with timely and accurate delivery date promises:

1. **Build-to-stock:** The products are stocked at and directly shipped from finished goods inventory. Customer service levels are determined based on a forecast of independent demand and statistically-based safety stock levels. Make-to-stock manufacturing environments offer customers the fastest delivery.

2. **Build-to-order:** The production process is initiated after the receipt of a customer order. Sufficient production capacity and flexibility are necessary prerequisites to appropriately schedule incoming orders. Performance to schedule is of paramount importance if correct customer order delivery date promises are going to be made. Build-to-order manufacturing environments offer customers the longest delivery times. Long lead time components may be stocked based on a forecast to reduce delivery time.

3. **Assemble-to-order:** The assemble-to-order production environment is a special case of the build-to-order production environment. Product subassemblies or options that are configured based on the customer's specification of the final product are stocked in anticipation of customer demand. The level at which product subassemblies or options are stocked is based on a forecast. Assemble-to-order manufacturing environments offer customers significantly faster delivery times than build-to-order manufacturing environments. A final assembly schedule is used to pull the subassemblies or options necessary to satisfy a particular customer order configuration. The final assembly process typically takes only a matter of hours, whereas build-to-order manufacturing environments may take days or weeks to deliver a product.

The three general types of production environments are shown in Fig. 2.6.2 and the relationship of the master production schedule to the final assembly schedule is shown for the various product structures.

The ability of customer order service to achieve high customer satisfaction levels is dependent upon its ability to review and modify product availability and scheduling on a *daily* basis. Greater MRP II system response times such as a week will place the company at a competitive disadvantage relative to a competitor who has daily or real-time system capabilities. What products are available-to-promise by customer order service is derived from an important master planning calculation. Delivery promises must be based on the current

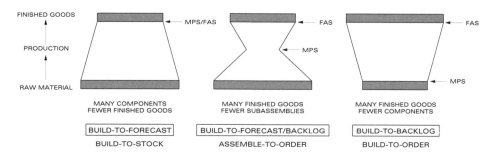

Figure 2.6.2 Product structure.

status of semifinished goods inventory (assemble-to-order), finished goods inventory (build-to-stock), or current capacity availability and schedule performance compared to plan (build-to-order). All of this must be considered while maintaining customer delivery date commitments already made for the future. Excellence in customer service is only achieved after the customer's order requirements are clearly understood and current manufacturing inventory, capacity, and schedule positions are accurately known.

The forecasting function of demand management is critical to the process of determining a company's inventory investment position (Fig. 2.6.1). High-mix, low-volume manufacturing environments present a special and difficult problem when calculating appropriate levels of inventory investment. Traditional forecasting methods such as moving averages and exponential smoothing are intended for items that have relatively smooth and continuous demand. High-mix, low-volume manufacturing environments exhibit demand patterns that are unusual. Periods of demand are followed by periods of zero demand and aggregate demand volumes are typically insufficient to produce particular products on a daily basis. Although most of the safety stock should be invested for those products representing the greatest sales volumes, statistically-based safety stock algorithms will place most of the safety stock for lower volume products. The reason for this is clear. The magnitude of forecast error is greater for the lower volume products. The laws of probability will incorrectly inflate inventory investment for a high-mix, low-volume manufacturing environment and damage a company's competitiveness in the marketplace. Using a statistically-based safety stock strategy for a high-mix, low-volume manufacturing environment is not a core competency.

Investment in safety stock is made for four reasons:

1. To cover forecast errors that are less than actual demand
2. To cover supplier deliveries that are late
3. To cover supplier deliveries that are less than the quantity ordered
4. To achieve a particular customer order line-fill rate.

To effectively set safety stock levels, the safety stock strategy must be inexorably linked to the inventory investment level developed during the production planning process that also provides linkage to the operating budget formulated during the business planning process.

Focus forecasting® is the preferred method of predicting future demand for a high-mix, low-volume manufacturing environment. Based on an eighteen-month actual demand history, a computer will look back three time periods that have already occurred and project these demands based on simple user defined strategies. Simple strategies are used—such as what was sold last year for the same three periods is projected to what will be sold this year, or what was sold in the previous three months to what will be sold in the next three

® *Registered trademark of Bernard T. Smith*

months. Buyers who are the primary users of the forecasting system are responsible for establishing the particular strategies, and fewer than 10 are sufficient to yield excellent results. The key is to keep the strategies simple so that they are easily understood. The strategy that performs the best for these three periods is the strategy that is used to forecast the future. The closer the ratio of forecast sales to actual sales is to one, the better the strategy. Seasonality, trends, bias, and alpha factors that are used in exponential smoothing are not a part of focus forecasting simulations. As a result, focus forecasting requires no file maintenance. When buyers are considering a buying decision, a code is placed on the purchase requisition under review indicating what forecasting strategy was used. New products are specially coded for appropriate buyer review and are also forecast based on available demand history from the most recent past. Buyers are also flagged to review buys if unusual demand patterns occur. Focus forecasting demands accurate demand history. To accomplish this, it will be necessary to track customer orders that were canceled as well as reordered to prevent the problem of inflated demand history.

The aggregate sales dollars are planned and tracked during production planning. The sales at cost are used to begin planning inventory levels. If sales differ significantly from budgeted sales, the inventory manager must follow the controller's budget until a new budget is released. For this reason, budgeted sales should be reviewed at least monthly. By referencing the relationship between line-fill rate and months on-hand for the past year(s), months on-hand for future budgeted sales can be appropriately derived to achieve desired line-fill rates. Line-fill rate is a measure of customer service that is based on the number of lines for different products that a customer receives relative to the total lines ordered, irrespective of the quantity ordered for a specific line item. Line-fill rates are dynamic and change from month to month. On-hand inventory at the end of one month that supports the following month's sales is not the same for each month due to the dynamics of changing order mix and seasonality. For this reason, the inventory position of the company is planned monthly during the production planning process.

Once the inventory manager has established the desired months on-hand inventory levels, the months on-hand inventory levels are converted to sales at cost and compared to the controller's budgeted sales at cost. If the controller foresees cash flow problems, the inventory manager will have to decrease planned inventory levels. If the company has a cash surplus, inventory levels may be increased to capture additional sales and positively affect profits for the desired period, providing there is adequate customer demand. For high-mix, low-volume manufacturing environments, raising inventory levels is a risky business and there are specific conditions that must typically occur before a decision is made to raise inventory levels over planned levels. Responding to a competitive threat such as competitor product promotions or other competitor moves such as shortening quoted delivery time are instances under which planned levels of inventory may be raised. Such information will permit the inventory manager to raise inventory levels for specific products based on rational reasons. Raising inventory levels on a high margin product to achieve fast delivery along with a price decrease is an effective way to catch a competitor sleeping and win in the marketplace if the competitor is unable to respond in time. Due to the volatility of external customer demand, raising inventory levels in a high-mix, low-volume manufacturing environment above planned levels should only occur for rational strategic reasons.

To effectively set safety stock levels, the inventory manager should list all products in order of nonincreasing demand and subdivide all items into at least six equal parts based on annual demand dollars. Each subdivision will represent the same revenue. Each month's supply of safety stock for each subdivision will, therefore, represent the same inventory cost. The inventory manager must allocate most of the safety stock to the highest volume product subdivision and allocate the least safety stock to lowest volume product subdivision. The inventory manager should adjust the month's supply for each subdivision to obtain the best overall line-fill rate. The total months of supply for all subdivisions in dollars should equal the budgeted inventory levels set by the production planning process.

Once safety stock levels are calculated it is important to determine what length of time should pass between buying decisions. This concept is referred to as the economic review frequency. The ideal review frequency reduces lines of work and carrying cost by a percentage of the total annual inventory investment. The economic review frequency (ERF) is calculated as follows:

$$ERF = \sqrt{\frac{.5(x)(annual\ volume)}{number\ of\ items}} \quad where: \quad x = \frac{Inventory\ carrying\ cost}{Average\ inventory\ investment} \qquad (2.6.7)$$

The inventory carrying cost is the cost to maintain the average inventory investment expressed in dollars and is equal to the total lines of work associated with the average inventory investment. The lines of work for each vendor is calculated by simply multiplying the number of items for each vendor by the review frequency associated with that particular vendor. The total of lines of work is the sum of the lines for all vendors. Once x is determined, the economic review frequency for a particular vendor is calculated using Eq. (2.6.7) after determining the vendor's annual volume expressed in dollars and the number of items ordered. Freight restrictions are considered by dividing the total annual volume in dollars for a particular vendor by the minimum freight restriction in dollars. This represents the maximum number of orders that can be placed to this vendor to receive free or discounted freight costs. In the case where volume discounts positively affect gross profit, the maximum order frequency is calculated by dividing the total annual volume in dollars for a particular vendor by the volume discount expressed in dollars. Freight and volume discount order frequency limits are important in the case where the economic order frequency calculation recommends an order frequency that is greater than required to obtain a particular discount.

The reorder point time is equal to the sum of the replenishment lead time, safety stock time, and review time. Vendor lead time performance is calculated based on the vendor's most recent delivery history and is the average lead time of the last three orders placed to the vendor. This lead time is used for all items in the vendor's line. The total quantity required is derived based on the forecast of the item over the reorder point time. The actual quantity ordered is equal to the reorder point quantity minus the on-hand quantity. The total inventory investment is equal to one-half the review time (expressed in dollars) plus the safety stock time (expressed in dollars) for a particular purchased item under consideration.

Focus forecasting is a holistic method of inventory management that is based entirely on simulation for anticipating future demand and links the safety stock strategy of a company to its operating budget. Focus forecasting has produced phenomenal results. Studies have consistently shown that focus forecasting significantly outperforms stochastic forecasting methods. For one particular company that used seven simulation strategies, 100,000 unique item forecasts were performed on a monthly basis, and forecast dollars over any three-month period deviated less than 1 percent from actual dollars during

Period	1	2	3	4	5	6
Forecast	50	50	50	50	50	50
Orders	62	44	33	25	39	0
PAB	8	164	131	81	31	181
ATP	8	59				
Cum. ATP	8	67	67	67	67	
MPS		200				200

Lead time: 0	Lot size: 200	Demand time fence: 3
On hand: 70	Safety stock: 0	Planning time fence: 5

Figure 2.6.3 Master planning tableau: build-to-order, assemble-to-order.

relatively stable years. For one particular three-month period, the difference between forecast dollars and actual dollars was 0.3 percent.[137,138]

The master planning module of Fig. 2.6.1 represents a disaggregation of the production plan into specific salable products that a company is going to manufacture. The master production schedule (MPS) answers the questions of what, how many of, and when each product is going to be produced. Based on focus forecasting simulations, forecast demand for each product is input by time period (e.g., weeks or days) to its associated master scheduling tableau as shown in Fig. 2.6.3. It is important to note that formats for master planning tableaus may vary considerably across different software packages.

The projected available balance (PAB) quantifies the expected inventory balance for the future and is calculated prior to and after the demand time fence as follows:

Before the demand time fence (execution):
PAB = on-hand balance or prior period PAB + MPS - customer orders   (2.6.8)

After the demand time fence (planning):   (2.6.9)
PAB = prior period PAB + MPS - customer orders or forecast (whichever is greater)

Prior to the demand time fence the forecast is consumed by actual customer orders that the company has committed resources to build and the forecast is ignored. The demand time fence often represents the backlog time horizon for build-to-order or assemble-to-order manufacturing environments. After the demand time fence and before the planning time fence, the company has not completely committed resources. During this period it is possible that additional customers' orders will arrive or scheduled orders may be canceled, thus altering the PAB. The greater of customer orders or the forecast is used to correctly assess what the PAB for this future time period will be. Period 1 of Fig. 2.6.3 represents the action bucket or "today," assuming that periods represent actual manufacturing days. Periods could represent weeks, and the action bucket would represent the first day of the week. After period 1 has elapsed, it is dropped and period 2 will subsequently represent the action bucket. Additionally, period 7 will become visible and thus, maintain the visibility of six total planning periods. In this manner, the master planning tableau will continuously move through time. Further visibility into the future can be displayed as required. Additional flexibility is afforded by the ability to define planning periods up to the planning time fence as days or weeks while defining planning periods beyond the planning time fence as days, weeks, months, or quarters. For high-mix, low-volume manufacturing environments where demand volatility is a way of life, daily visibility is essential within the planning time fence. Daily planning facilitates timely response to change with accurate placement of planned order releases and planned order receipts. For daily planning, a manufacturing day calendar is used that consecutively numbers "actual" manufacturing days worked. Plant shut-downs, holidays, or other nonworking days are excluded from the manufacturing day calendar (a.k.a. M-day calendar). If lead times occur over a period of hours or days, they will not be artificially inflated due to weekly planning periods.

Time fencing is critical. The planning time fence represents the total planning horizon and is at least as long as the total cumulative lead time to produce the product including the time to procure all required material. The demand time fence represents the time required to assemble the product and may or may not include component assembly time. Customer orders within the planning time fence are *firm planned orders*. Firm planned orders override the computer system and prevent it from making changes to the material requirements plan. Changes can only be performed manually. There are good reasons for this. Increasing an order within the demand time fence may create a production overload and cause other orders to be late. This may create the need for overtime. The cancellation of an order will create the additional cost associated with the waste of excess capacity at the capacity constraint operation. Changes within the demand time fence create additional cost and the risk of late deliveries and must only be permitted after the approval of top management. The adverse consequences associated with changes after the demand time fence and before the planning time fence may be less serious; because manufacturing resources are not yet committed. Increasing a particular order (expediting) may be counterbalanced by the canceling of another similar order (de-expediting) or the decision to outsource if the order is sufficiently large. When sufficient material is not available to increase an order, purchasing may be able to expedite delivery. While there is some flexibility to facilitate change in order mix, all requested order changes may not be feasible. Within the planning time fence, manufacturing may have to reject specific order change requests. Master schedules must not be changed at a rate faster than manufacturing resources or suppliers can cope with. Management must have a clearly defined policy that states the level of management approval required to accept order change requests. Beyond the planning time fence the computer system should control all planning logic.

The available-to-promise (ATP) calculation is of paramount importance. Available-to-promise projects the future inventory or capacity that customer order service may promise to the customer. The calculation of available-to-promise is as follows:

ATP = on-hand balance (action bucket only) + MPS - sum of customer orders   (2.6.10)
before the next MPS

The cumulative ATP is shown by time period up to the planning time fence in Fig. 2.6.3. If ATP were to go negative, it would be set to zero, and the ATP previous to the negative ATP would be decremented by the negative ATP value. A lot-for-lot (L4L) lot sizing technique would set the MPS equal to the net customer order requirement. Lot-for-lot is the preferred choice for high-mix, low-volume manufacturing environments due to the reduction of inventory levels and improved manufacturing cycle time that results.

Under the condition of level and continous annual demand, lot sizes can be based on the concept of the economic order quantity (EOQ). The EOQ is calculated by determining the minimum total cost associated with the interaction of ordering cost and inventory carrying cost for a particular item under consideration. A decrease in the quantity ordered of a particular item will result in reduced carrying cost. A decrease in the amortization of setup cost will also occur. Conversely, an increase in the quantity ordered of a particular item will reduce the annual ordering cost and improve the amortization of setup cost. Although lot sizes based on the EOQ are subject to the restriction of continuous annual demand, the utility of EOQ as a management inventory control tool is based on its ability to guide the management decision-making process by defining the relationships in connection with ordering cost, setup cost amortization, and inventory carrying cost.

If the master planned items are make-to-order or assemble-to-order, a final assembly schedule will be required to specify the desired customer configuration as well as timing for delivery. For master planned items that are make-to-stock, the MPS and final assembly schedule are one and the same. Safety stock should rarely be carried for build-to-order products. Customers will normally place orders ahead of the total cumulative time required to acquire materials for producing and delivering the product. For build-to-stock manufacturing environments, customer orders and available-to-promise are not relevant. The master production schedule is developed to ensure that a predetermined customer service level for off-the-shelf delivery is maintained by using the forecast: plus safety stock to buffer incoming customer order volatility.

Period	1	2	3	4	5	6
Forecast	50	50	50	50	50	50
PAB	20	70	20	70	20	70
MPS		100		100		100

Lead time: 0   Lot size: 100   Demand time fence: 3
On hand: 80   Safety stock: 10   Planning time fence: 5

Figure 2.6.4 Master planning tableau: build-to-stock.

The master planning tableau for a build-to-stock manufacturing environment is shown in Fig. 2.6.4.

The projected available balance quantifies the expected inventory balance for the future and is based solely on the forecast demand. The projected available balance is calculated prior to and after the demand time fence as follows:

Before the demand time fence (execution):

PAB = Prior period or on-hand balance + MPS - forecast   (2.6.11)

After the demand time fence (planning):   (2.6.12)

PAB = Prior period PAB + MPS - forecast or customer orders (whichever is greater)

Customer orders are included only in the case where unusually high demand results in a stockout where a backlog has developed or the risk of a stockout is imminent. Beyond the demand time fence it is possible to respond to demands in excess of forecast demand due to carrying safety stocks of component inventory. During the monthly production planning meeting it will be important to review and possibly modify the forecast and safety stock levels if the risk of poor customer service is rationally established (e.g., via marketing or sales information) as likely. It is possible that the demand in excess of the forecast plus safety stock level is an aberration.

The overriding goal of master planning is to develop a plan that can be executed at least 95 percent of the time. This must be accomplished in a way that is profitable to the company and achieves good customer service. Rough-cut capacity planning (RCCP) is a must if we are to ensure that the master plan is valid (i.e., doable) (Fig. 2.6.1). Rough-cut capacity planning focuses on the detailed day-to-day schedules at the constraint operations. By assuring that temporal constraints are properly managed, the RCCP predicts the master production plan's ability to achieve the company's goals. A bill of resources is developed and represents the various workcenter operations through which the products will flow and includes the standard assembly times for each unit produced at each operation. The bill of resources is the time to produce only one unit of each product. Multiplying the number of units produced by the resources required per unit establishes the total time required at each workcenter operation. This process is performed by time period and is offset by the required lead time to obtain the start date for each product. Once the time phased RCCP is developed, the total resource requirement at each workcenter operation is multiplied by an efficiency factor, and the aggregate capacity required by time period is compared to available capacity in bar chart format. Bar charts facilitate easy identification of resource overloads and underloads by time period. Periods of overloading or underloading may necessitate changes in the master production schedule. Validation of the master production schedule through RCCP is an iterative process. After a master production schedule is validated by RCCP, the master planner authorizes the material requirements plan explosion process.

The material requirements plan (MRP) module of Fig. 2.6.1 represents the disaggregation of the master schedule into a schedule for factory production and purchase orders necessary to achieve the master production schedule. The MRP process is guided by the bill of material, inventory data, and planning factors. The bill of material (BOM) is the complete product structure for the master scheduled end product (see Fig. 2.6.5). The BOM lists all purchased and manufactured items that make up the end product in the order that they are assembled and is developed by the research and development engineering department of the company. Inventory data includes on-hand inventory balances, lead times, safety stock, lot sizes, allocations, scrap factors, and the status of open orders by due date and quantity. Planning factors define time periods, planning horizon, and replanning frequency. Planning factor data define the boundaries of how the MRP process is controlled.

MRP has two modes of replanning frequency. The regenerative mode typically rebalances all computer system records on a weekly basis and time periods represent one week. The net change mode rebalances records when necessary and maintains a perpetually up-to-date material requirements plan. Time periods represent one day and are sequenced based on a manufacturing day calendar. Regenerative systems group multiple days into each time period and such time periods are often referred to as time buckets. The net change mode of operation is often referred to as bucketless as contrasted to regenerative mode MRP systems. For high-mix, low-volume manufacturing environments, it is essential to use the net change mode of operation. The net change mode of operation facilitates increased responsiveness to change and reduced cumulative lead times. When lead times are hours or days, the regenerative mode of operation will artificially inflate lead times to a minimum of

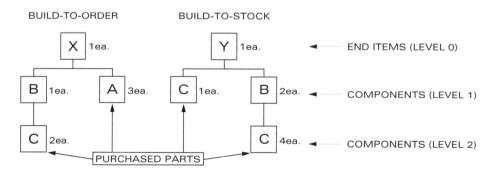

Figure 2.6.5 Bill of material product structure.

PRODUCT X-MPS

Period	1	2	3	4	5	6
Forecast	50	50	50	50	50	50
Orders	62	44	33	25	39	0
PAB	8	164	131	81	31	181
ATP	8	59				
Cum. ATP	8	67	67	67	67	
MPS		200				200

Lead time: 0   Lot size: 200   Demand time fence: 3
On hand: 70   Safety stock: 0   Planning time fence: 5

PRODUCT Y-MPS

Period	1	2	3	4	5	6
Forecast	50	50	50	50	50	50
PAB	20	70	20	70	20	70
MPS		100		100		100

Lead time: 0   Lot size: 100   Demand time fence: 3
On hand: 80   Safety stock: 10   Planning time fence: 5

COMPONENT A-MRP

Period	1	2	3	4	5	6
Gross requirements		600				600
Scheduled receipts		1000				
Projected on-hand	300	700	700	700	700	100
Projected available	300	700	700	700	700	100
Planned order receipts						
Planned order releases						

Lead time: 1 Period   Lot size: 1000
On hand: 350   Safety stock: 50

Figure 2.6.6 Material requirements planning.

COMPONENT B-MRP

Period	1	2	3	4	5	6
Gross requirements		400		200		400
Scheduled receipts		1000				
Projected on-hand	0	600	600	400	400	0
Projected available	0	600	600	400	400	1000
Planned order receipts						1000
Planned order releases				1000		

Lead time: 2 periods   Lot size: 1000
On hand: 200   Safety stock: 200

COMPONENT C-MRP

Period	1	2	3	4	5	6
Gross requirements		100		2100		100
Scheduled receipts						
Projected on-hand	500	400	400	-1700	-1700	-1800
Projected available	500	400	400	4300	4300	4200
Planned order receipts				6000		
Planned order releases		6000				

Lead time: 2 periods   Lot size: 6000
On hand: 550   Safety stock: 50

Figure 2.6.6 (*continued*) Material requirements planning.

one week. A high-mix, low-volume manufacturing environment that uses the net change mode has a competitive advantage in responsiveness over a less adept competitor that uses the regenerative mode. In today's competitive marketplace, no company has the luxury of giving away a competitive advantage in customer responsiveness.

Consider the product structures for products X and Y as shown in Fig. 2.6.5. Based on the master production schedules, product structures, inventory data, and planning factors, the material requirements plan is developed as indicated in Fig. 2.6.6.

The gross requirements for component A are dependent upon the master production schedule and product structure for product X. A quantity of 200 product Xs is required in period 2. From the product structure for product X (Fig. 2.6.5), 3 As are required for each unit of product X. Therefore, the gross requirement for component A in period 2 is 600 *[200X (3ea. A)]*. Component B's gross requirement is dependent upon both product X's and product Y's master schedules and bill of material (BOM) product structures. One each of component B is required for each product X and 2 each are required for product Y. Based on the period 2 master production schedules for products X and Y, the net requirement is 200 of product X and 100 of product Y. The total gross requirement for component B in period 2 is calculated as 200X(1ea. B) + 100Y(2ea. B) = 400. The gross requirement for component C is dependent upon the net requirement for product Y and component B as indicated on the product structures for products X and Y (Fig. 2.6.5). A net requirement (i.e., planned order release) for 1,000 units of component B occurs during period 4 of component B's MRP explosion, and a net requirement for 100 units of product Y occurs during periods 2, 4, and 6. The gross requirement for component C in period 4 is calculated as 100Y(1ea. C) + 1,000B (2ea. C) = 2,100. The gross-to-net logic of MRP is performed by time period for each component using the bill of material product structure as a guide. The gross-to-net logic only occurs at the lowest level for the particular component under consideration within the product structure. The upper level parent for which a particular component is dependent must be calculated first. Dependencies within the product structure are sometimes referred to as *precedent constraints*. It is essential to have 100 percent accurate bill of material product structures for all products produced.

Safety stocks of component parts are maintained if the component part has independent demand, for example, as would occur for service center parts replacement orders. The on-hand balance at the beginning of the first period is adjusted by subtracting safety stock. Projected on-hand balances are calculated using the following equation:

Projected on-hand = on-hand at beginning of period + scheduled receipts - gross requirements
$$(2.6.13)$$

The projected on-hand does not take into consideration planned orders for items that have yet to be released. Only after a planned order is released during the action bucket will the planned order receipt become a scheduled receipt on the MRP tableau. A planned order receipt is indicated for the time period that a shortage will occur. Shortages are indicated when the on-hand balance becomes equal to or less than the safety stock level. The planned receipt is equal to the lot size rule in effect. The planned order release that is associated with the planned order receipt is scheduled backwards in time for the number of periods equal to the lead time required to produce or purchase the required number of items. The projected available balance takes into consideration planned order receipts and will never become negative. Projected available balances are calculated using Eq. (2.6.13) and adding planned order receipts. If performance to lead time is late, the lead time offset may be increased the appropriate amount of time required to buffer against uncertainties in delivery time.

The requirement for accurate on-hand inventory item count balances should be obvious at this point. The requirement for accurate on-hand inventory item count balances must be management policy. Failure to maintain accurate inventory item count balances will corrupt the entire MRP explosion process and the objectives of the company will not be achieved. The dollar value of inventory does not assure that item counts are accurate and must not be used as a measure of item count accuracy. Inventory item count accuracy must be a minimum of 95 percent accurate for each and every item and is a *prerequisite* to using MRP successfully. Once planned orders are released, it will take time to pull the required material and supply it to the appropriate operation. For this reason, material must be allocated once an order is released. Allocations of this nature are often referred to as *uncashed requisitions*. Allocated material must not be considered in the MRP planning logic. After material is removed from a stock location (i.e., when the allocated material is actually used), the on-hand and allocated balances are reduced accordingly. A process called cycle counting must be instituted where cycle count personnel are continuously employed to verify stock location item count accuracy. If item count accuracy errors are discovered, the item count must be readjusted. The root cause must be identified to prevent the recurrence of problems in the future. A companywide education and training program in inventory management basic fundamentals is required to successfully maintain accurate inventory records. The work force must understand why inventory item count accuracy is a must in addition to what and how accurate item count records are maintained.

At this point, capacity requirements planning (CRP) is performed to validate the tentative material requirements plan (Fig. 2.6.1). CRP offers significantly improved resolution over rough-cut capacity planning. The routing file details process step by process step: how the particular product is manufactured, run time per unit, setup time per batch, and interoperation move times. The workcenter file details each workcenter in terms of number of machines or workstations, number of shifts, queue time, and available capacity. MRP has calculated planned order releases and planned order receipts as well as scheduled receipts by time period and quantity. The routing file is used to direct the CRP process. The manufacturing lead time is the sum of

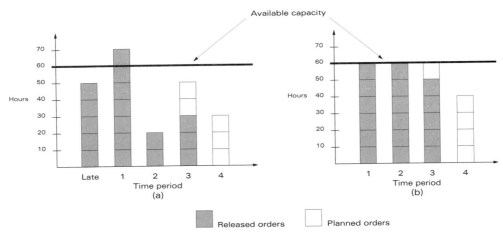

Figure 2.6.7 Workcenter load profile.

the queue, setup, run, wait, and move time required to produce the product. Queue, wait, and move times should be reduced to the point that only the setup and run time elements of manufacturing lead time are used. To create the capacity load profile, the planned and scheduled receipts are backwards-scheduled in time based on the manufacturing lead time to arrive at the latest start date. By forward-scheduling the manufacturing lead time from the MRP planned order release date, the earliest completion date can be ascertained. A workcenter load profile can thus be obtained and is usually in bar chart format as indicated by Fig. 2.6.7.

The workcenter capacity load profile will often indicate periods of overload and underload due to the fact that MRP loads workcenters assuming infinite capacity (Fig. 2.6.7a). Based on information contained in the CRP profile, the MRP planned order releases and planned order receipts may be adjusted to alleviate periods of underload or overload. Consider Fig. 2.6.7b. All late work is moved to the first period. All work originally scheduled for period one is de-expedited to period two with the exception of 10 hours of work. Twenty hours of work originally scheduled in period two is de-expedited to period three and 10 hours of work planned for release in period three is rescheduled to period four. Based on these changes, the resultant load profile will level and the MRP plan will be validated. This process is iterative and must occur if the objectives of the company are going to be achieved. If changes to the MRP plan are made resulting in late orders occurring due to a correction made to an on-hand balance or other irreconcilable problems, the MRP system will generate action messages notifying the master planner of the problem. The master planner should make every attempt to solve the problem at the MRP level and only as a last resort should the master planner change the master plan. There are several software packages available that will effectively perform CRP. The preferred method of capacity requirements planning in a high-mix, low-volume environment is finite capacity planning. Finite capacity planning does not assume infinite capacity and loads the factory based on mathematical models without creating overloads. Finite capacity planning is also known as operations sequencing and is addressed in the next section (2.7 Scheduling).

Once the material requirements plan is approved, factory and purchasing execution may begin (Fig. 2.6.1). For effective purchasing execution, the MRP plan for purchased parts A and C (Fig. 2.6.6) should be shared with the appropriate vendors for several periods beyond the planning time fence. This will allow the vendor to respond to any anticipated changes much more effectively. Vendors will be able to act upon change proactively rather than reactively, when they receive purchased part requirements for the future updated on a monthly basis. If a purchasing problem occurs, the buyer must communicate this to the master planner as soon as possible to allow the master planner as much time as possible to evaluate, recommend, and implement the most appropriate remedy. If the process of vendor communication uses the ANSI X-12 standard for electronic data interchange (EDI), the efficiency and effectiveness of the purchasing function will be enhanced. The electronic sharing of planning system information is becoming common practice. By removing buyers from the day-to-day vendor communication loop they can better maintain and establish vendor relationships and contracts. The advantages of EDI can be realized up and down the supply chain.

During factory execution, it is critical to monitor factory performance to plan through a process called input-output control. Based on realistic schedules, the factory is responsible for the on-time delivery of build-to-order and assemble-to-order products as well as the line-fill rate for build-to-stock products. To achieve on-time delivery and line-fill rate goals, the greatest concern is work-in-process (WIP) inventory. WIP control facilitates smooth flow at minimum inventory levels. Less than optimum scrap and yield levels have an adverse impact on the factory's performance objectives. To ensure that the master schedule is achieved, gross requirements for component parts will have to be inflated by the percentage scrap and defect levels encountered. The gross requirements will be multiplied by the planning factor: *1 + scrap/(total material)+ defects/(total material)*. Continuous quality improvement in process and material, as well as a sound preventive maintenance program for machinery, will reduce cost and improve the performance of any factory. Each workcenter should monitor its performance using an input-output control report similar to the one found in Fig. 2.6.8.

The actual deviation for input and output is the actual hours minus the

INPUT

Weekly Period	1	2	3	4	5	6
Planned hours	540	525	550	520	550	530
Actual hours	550	500	560	510	525	540
Cumulative deviation*	10	-15	-5	-15	-40	-30

OUTPUT

Weekly Period	1	2	3	4	5	6
Planned hours	530	530	530	530	530	530
Actual hours	540	525	540	545	510	540
Cumulative deviation*	10	5	15	30	10	20

QUEUE

Weekly Period	1	2	3	4	5	6
Planned	490	485	505	495	515	515
Actual	490	465	485	450	465	465

Beginning queue: 480 hours

*Tolerance: 50 hours

Figure 2.6.8 Input-output control report.

planned hours. The planned and actual queue are calculated for each time period using the following equations:

Planned queue = prior period queue + planned input - planned output   (2.6.14)

Actual queue = prior period queue + actual input - actual output   (2.6.15)

Based on the results of Fig. 2.6.8, the queue was less than planned due to actual output being greater than planned. In reality, queues must never be planned for, but their occurrence must be closely monitored. The use of generic Kanban guarantees that queues will never accumulate out of control and thus, will minimize work-in-process inventory levels. This results in short manufacturing lead times that translate into improved customer responsiveness. The input-output control report provides excellent visibility to what is happening on the factory floor. When significant deviations occur, corrective actions can be taken. Significant deviations may occur for any number of reasons (e.g., machine breakdowns, stockouts, etc.) and should be *immediately* reported to the master planner.

Effectiveness of master schedule performance is not only based on how well it was set but on how well it is followed and executed. MRP II is a closed-loop system that integrates all functional areas of the business to achieve the company's mission and specific business objectives. MRP II effectively links the company to the customer and marketplace. One of the shortcomings of MRP is the requirement for establishing lead times that are constant when in fact they are dynamic variables. Fixed lead times create shop floor scheduling and priority problems. Another problem is that capacity requirements planning is not integrated with MRP and rough-cut capacity planning is not integrated with the master plan. Attempts to manipulate the master plan or material requirements plan to alleviate overloads often prove to be difficult. A change that is made to solve an overload problem at one workcenter may cause an overload at another. What is needed is a capacity planning process that will detect overloads and underloads during the MRP process and automatically reschedule the MRP system in an integrated way to effectively use available factory capacity. To achieve maximum benefit from MRP, management and users must be thoroughly educated in its capabilities and shortcomings. Educated users who are led by highly competent managers can create substantial operating improvements that offer a competitive advantage over less educated competitors. In fact, education may be the only real lasting source of competitive advantage in today's competitive global marketplace.

2.7 Scheduling

The primary objectives of scheduling for a high-mix, low-volume manufacturing environment are as follows:

- On-time delivery
- Minimum total processing time
- Maximum resource utilization
- Minimum inventory.

In reality, these objectives are in conflict with each other. Consider Fig. 2.7.1. Increased resource utilization results in a consequent reduction in due-date performance (i.e., on-time delivery) due to the use of large production batches required to avoid machine setups. Conversely, if machine setups are maximized to reduce cycle time (i.e., total processing time), a consequent reduction in resource utilization will occur. The controlling variable is lot size. The lot size determines the amount of work-in-process (WIP). The

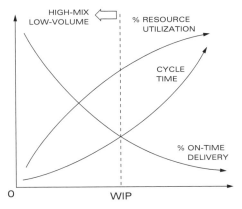

Figure 2.7.1 Tactical competitive positioning.

design of a production schedule must be based on the competitive objectives set forth in the business plan. For high-mix, low-volume manufacturing environments, competitive advantage should be pursued in the areas of quality, responsiveness, and on-time delivery. To achieve competitive advantage, the schedule must be based on relatively small lot sizes as compared to the competition. In order to produce the small lot sizes required, sufficient capacity must exist to facilitate the large number of machine setups required. A high level of resource utilization is indicative of low-mix, high-volume manufacturing environments that are competing on price. Maximizing resource utilization in a high-mix, low-volume manufacturing environment is not a core competency.

Scheduling methods are based on models that represent approximately the production process. Scheduling methods address two important issues:

1. **Allocation:** assigning manpower and/or machines to particular jobs.
2. **Sequencing:** the order in which given jobs are to be performed.

The two general classifications of scheduling methods are static and dynamic. For static scheduling problems a finite amount of work is performed. The workcenters initially may be loaded or empty but are empty at the conclusion of the study. Performance criteria are based on the total time to process all jobs (i.e., makespan). While optimization can be achieved for a limited set of problems that are small in scope, heuristic methods are used for larger problems using deterministic or stochastic processing time data. Dynamic scheduling problems are based on steady-state system behavior when a continuous supply of jobs is available to be processed. Research has focused considerably on stochastic system behavior using a variety of dispatching rules. Queueing models have also been developed. Measures of sys-

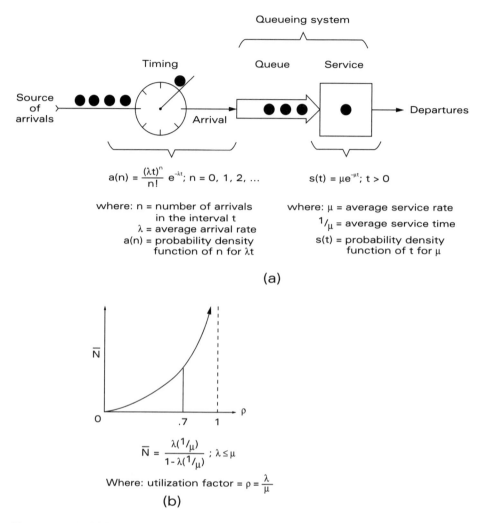

where: n = number of arrivals
in the interval t
λ = average arrival rate
a(n) = probability density
function of n for λt

$$a(n) = \frac{(\lambda t)^n}{n!} \, e^{-\lambda t}; \; n = 0, 1, 2, \ldots$$

where: μ = average service rate
$1/\mu$ = average service time
s(t) = probability density
function of t for μ

$$s(t) = \mu e^{-\mu t}; \; t > 0$$

(a)

$$\overline{N} = \frac{\lambda(1/\mu)}{1 - \lambda(1/\mu)} \; ; \; \lambda \leq \mu$$

Where: utilization factor $= \rho = \frac{\lambda}{\mu}$

(b)

Figure 2.7.2 M/M/1 queue.

tem performance based on workcenter utilization, average work-in-process, average flow time, number of jobs in the system, average lateness, lateness variance, and several other measurements are used as figures of merit for overall dynamic system performance. The development of a schedule is limited by available capacity and temporal activity precedent constraints. A temporal activity precedent constraint is the sequence in which operations must be performed and is also known as the process routing.

Queue, load, and work-in-process are terms often used to denote problems associated with waiting lines. A queueing system is defined as any system in which arrivals place demands on a resource that has finite capacity. One of the most important nontrivial queueing models is the celebrated M/M/1 queue.[52] The M/M/1 queue is Markovian in that interarrival times and service times are independent. A Markovian system is memoryless. The interarrival time is Poisson distributed (discrete) and the service time is exponentially distributed (continuous). The general structure of a waiting line is depicted in Fig. 2.7.2a.

Inexhaustible arrivals originate from a customer source. The queue and servicing operation together are referred to as a queueing system. The resulting behavior of the M/M/1 queue is counterintuitive. If the average arrival time is one unit every two minutes and the average service time is two minutes per unit, the queue will grow geometrically to infinity as shown in Fig. 2.7.2b. There is an extreme penalty for running the system near or at capacity. This type of behavior is characteristic of almost every queueing system one can encounter. The unbounded behavior which occurs as $\rho \to 1$ is due to the variability in both the interarrival time and service time. A reduction in the variance of either variable will reduce the average number of jobs in the system. The use of the master planning backward-scheduling technique (i.e., from the last operation to the first operation) will impose loads on upstream workcenters at random intervals even though the master schedule is reasonably leveled. Random arrivals will cause the practical capacity of a critical resource to gravitate to 70 percent. The load on constraint resources must be leveled through forward-scheduling (i.e., from the first operation to the last operation) so that work will arrive in a nonrandom fashion. Forward-scheduling determines the earliest completion date while backward-scheduling determines the latest start date, and a combination of both is required to effectively respond to changes. The theoretical basis for using Kanban as a mechanism to control interarrival time is derived from this queueing model. Service time variation may be reduced through worker training, machine preventive maintenance, and continuous quality improvement. A company that makes a continuous and concerted effort aimed at the reduction of interarrival time and service time variation will effectively increase its ability to efficiently use available capacity at constraint operations. This can provide a competitive cost advantage. Clearly, a competent production management team should exhaust all available resources to improve the utilization of existing constraint capacity before a decision is made to augment it.

The reduction of variance for interarrival times as well as service times at capacity constraint operations is critical to achieving competitive advantage. Significant reductions in variability will improve the predictability of production operations that will translate into excellence in customer order promising performance. Automation is commonly used to reduce the variability associated with interarrival and service times. The significance of reducing variation is the resultant ability to glean insights from production operations modeled deterministically. Although production operations will always exhib-

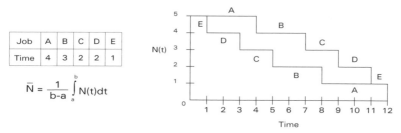

Figure 2.7.3   Shortest processing time.

it variability, deterministic models may be used to make inferences with regard to system behavior on average. Deterministic models have been developed for a variety of production operations scenarios from which specific methods of scheduling have emerged. Scheduling is primarily concerned with resource allocation and sequencing decisions for specific jobs to be produced. The time it takes to perform each operation must be quantified and the routing configuration (i.e., precedent constraints) of available resources must be defined. Resources are generally referred to as machines—although the resource may be people—and each discrete grouping of tasks (i.e., lot) to be processed is referred to as a job.

The fundamental building block of scheduling is the single machine sequencing problem. For all possible sequences of *n* given jobs *(n!)*, the makespan (i.e., total time to process all jobs) is a constant for the single machine sequencing problem. One criterion for optimization of the single machine sequencing problem is the mean flow time. By minimizing the mean flow time, inventory (the mean number of jobs in the system) is also minimized. For high-mix, low-volume manufacturing environments where there are several jobs with unequal work content times, the optimal sequence is based on the concept of shortest processing time (SPT). If a total of *n* jobs are to be processed, the mean number of jobs in the system may be mathematically expressed as shown in Fig. 2.7.3. Although the total number of possible sequences for *n* jobs is *n!*, the optimal sequence is arrived at by sequencing all jobs in nondecreasing order of processing time. This will minimize the area under the *N(t)* function. For the case of Fig. 2.7.3, the optimal sequence is E, D, C, B, A. The mean time that all jobs spend in the system for sequence A, B, C, D, E is *(4+7+9+11+12) / 5 = 8.6*. For sequence E, D, C, B, A, the mean time spent in the system is *(1+3+5+8+12) / 5 = 5.8*. The SPT rule is an important result. It can be demonstrated mathematically that SPT is optimal for the following criteria, irrespective of due date:[129]

- Average total time per job
- Average job lateness
- Average number of jobs in the system.

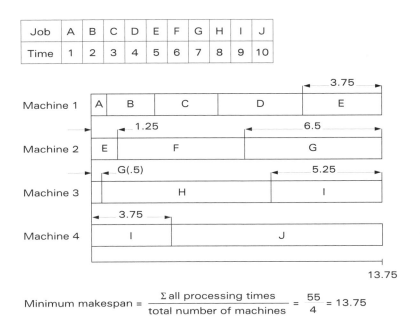

Job	A	B	C	D	E	F	G	H	I	J
Time	1	2	3	4	5	6	7	8	9	10

$$\text{Minimum makespan} = \frac{\Sigma \text{ all processing times}}{\text{total number of machines}} = \frac{55}{4} = 13.75$$

Figure 2.7.4 Parallel machine optimization.

If the optimization criterion is based on minimizing the maximum job lateness or minimizing the lateness variance, then running the jobs in due date sequence is optimal. Jobs may also be weighted by the inventory cost they represent. In this case, the mean flow time and mean inventory are minimized when the jobs are ordered in nondecreasing order of weighted inventory value. Under this scenario, the mean number of jobs may not be minimized. This methodology is referred to as weighted shortest processing time (WSPT). For complex system models where optimization by analytical methods is untenable, heuristics are employed and performance criteria are evaluated through simulation. SPT is the fundamental rule to which all heuristics are compared.

In the case of parallel machine models, the optimization of makespan can be achieved when preemption (i.e., lot splitting) of a job is permitted. McNaughton[94] developed an algorithm for optimizing makespan for the parallel machine problem when preemption is permitted and this is shown in Fig. 2.7.4. This procedure would have to be altered if setup time were involved. No direct solution to the parallel machine problem has been offered when job preemption is not permitted. Schedules can be constructed based on heuristics, and the schedules will almost always be nearly optimum for mean flow time if the load is divided as evenly as possible across all machines and SPT sequencing is subsequently performed for each machine. The parallel machine problem proves to be formidable. If *n* jobs are processed by *m*

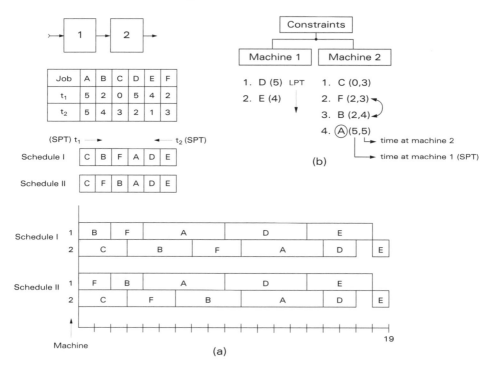

Figure 2.7.5 Two machine problem.

machines, the total number of possible sequences is equal to $(n!)^m$. Although each unique sequence must be evaluated in order to verify that a particular schedule is optimal, certain theoretical mathematical properties will allow a reduction in the total number of possible sequences for a limited number of cases.

Johnson[75] first offered an algorithm to optimize makespan for the serial two-machine problem. Consider Fig. 2.7.5a. To create the optimal schedule, the minimum processing time is determined. If the job is associated with machine one, it is placed in the first available position of the sequence. If the minimum processing time is associated with machine two, the job is placed in the last available position of the sequence. For equal processing times, the placement (i.e., ordering) of jobs is arbitrary. This process is repeated for each job. Since jobs B and F have equal times based on their processing times at machine one, their positions can be arbitrarily selected. Thus, there are two optimal schedules. All other schedules will increase the makespan (i.e., total time to process all jobs). The Gantt charts are developed for each optimal schedule and the makespan is 19.

There is a unique way to model the two-machine problem. Consider Fig. 2.7.5b and the following steps:

Step 1: Associate each job with its respective constraint. If any particular job processing time is greater at machine one, it is constrained at machine one; and if the job processing time is greater at machine two, it is constrained at machine two. In the case of job A (circled), it is equally constrained at machine one and machine two. The rule is to place this job with the most downstream constraint (machine two).

Step 2: The jobs that are constrained at machine one are placed in nonincreasing order of their processing time at machine one. This is referred to as the longest processing time (LPT) sequence. The ordering of jobs with equal processing times is arbitrary. Based on their processing times at machine one, the jobs that are constrained at machine two are placed in nondecreasing order. This is referred to as the shortest processing time (SPT) sequence. The sequence is arbitrary for jobs at machine two with equal constraint one processing times. Such is the case for jobs F and B. While the order of jobs at constraint two is indicated in Fig. 2.7.5b as C, F, B, A, it could also be C, B, F, A.

Step 3: The optimal sequence is obtained by first processing jobs in order at the most downstream constraint (machine two) followed by the processing of jobs in order at the gating constraint (machine one).

This algorithm offers some valuable insights. Any intraconstraint pairwise interchange of jobs (e.g., D↔E) will always result in a schedule that is not worse than any interconstraint pairwise interchange of jobs (e.g., D↔A). Thus, it is essential to sequence jobs based on constraint considerations.

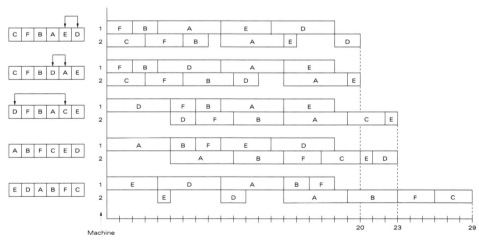

Figure 2.7.6 Sequence with job interchanges.

Consider the Gantt charts developed for a variety of intraconstraint and inter-constraint interchanges in Fig. 2.7.6. This conclusion makes intuitive sense when we consider that the worst possible schedule, E, D, A, B, F, C, is the inverse of the optimal schedule and requires a complete interconstraint interchange of jobs. For complex high-mix, low-volume manufacturing environments for which there is no suitable analytical solution, a production schedule can be developed using the following algorithm.

2.7.1 Multiple Constraint Synchronization (MCS) Algorithm:

OBJECTIVE: Minimize makespan.

Step 1:   Determine the job with the maximum processing time at the most downstream machine. Place this job in the first available position in the sequence.

Step 2:   Remove this job and its associated machine from consideration and return to step 1. If jobs or machines = 0, then stop.

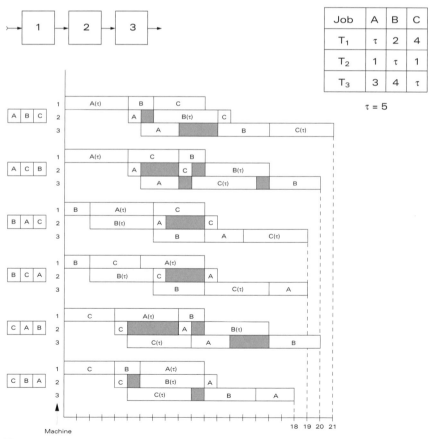

Figure 2.7.7 Three machine problem.

Job	A	B	C	D
T_1	τ	3	1	2
T_2	2	2	3	τ
T_3	1	τ	2	3
T_4	3	1	τ	1

$\tau = 4$

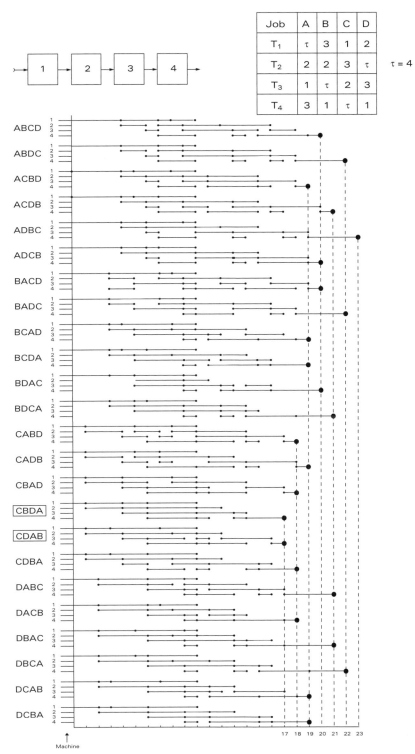

Figure 2.7.8 Four machine problem.

This algorithm is remarkably simple and will always find an optimal sequence for minimizing makespan for the multiple constraint serial machine problem. Consider the 3-machine multiple constraint serial machine problem of Fig. 2.7.7. The optimal sequence is C, B, A with a makespan of 18. The inverse of the optimal sequence is the sequence A, B, C that has the longest makespan (21).

Consider Fig. 2.7.8. While the MCS algorithm will find an optimal sequence it will not find them all. There are two optimal schedules for the 4-machine multiple constraint serial machine problem, and the optimal schedule C, D, A, B will not be determined using the MCS algorithm. It is interesting to note that there is more slack time for schedule C, D, A, B than the MCS schedule (C, B, D, A) yet both are optimal. Although schedules D, A, C, B and D, C, B, A have the least slack time for all possible sequences, they are suboptimal. Although the constraint times (τ) are equal for each job for the examples given in Fig. 2.7.7 and Fig. 2.7.8, they can be unique and not affect the MCS algorithm's ability to determine an optimal sequence. There is an advantage in reduced queue (i.e., WIP) if the constraint times for each job are equal. The optimal MCS sequences from Figs. 2.7.7 and 2.7.8 are shown in Fig. 2.7.9a.

Note that all constraints are aligned as closely as possible with respect to time (i.e., synchronized) for each optimal sequence. If the constraint times for each job are not equal, the optimal sequences are as shown in Fig. 2.7.9b. The most downstream constraints are constrained twice as long (2τ) as their associated upstream constraints. The constraint that feeds the most downstream constraint will complete its job(s) before the most downstream constraint completes its job. This results in an increase in the mean number of jobs in the system (i.e., WIP). The jobs in the queue, along with their associated processing times at their respective 2τ-constraint operations (indicated in brackets), are shown in Fig. 2.7.9b. The lot sizes for jobs should be selected in a manner that most closely equates all constraint times. The combining of dissimilar units of production within a particular defined job is a technique that may be used to facilitate the creation of equal constraint processing times and thus prevent a buildup of queue at the bottleneck resource. The minimum lot size is based on an integer multiple of the maximum constraint processing time per unit at the bottleneck operation. It is essential to ensure that the bottleneck operation is not overloaded and the minimum possible lot sizes are used to achieve balanced constraint times.

For serial flow, high-mix, low-volume manufacturing environments, the flow of work is unidirectional. Serial flow shops have a gating process that performs only the first operation and a terminal process that performs only

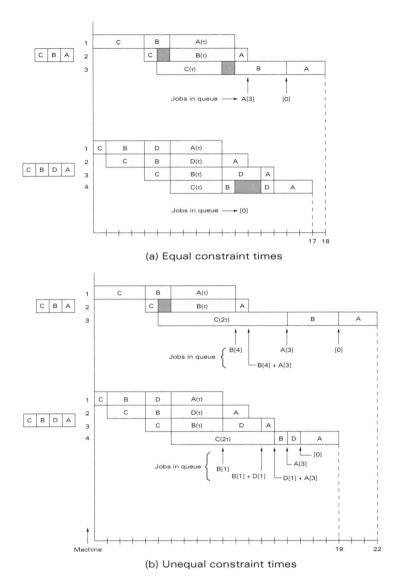

(a) Equal constraint times

(b) Unequal constraint times

Figure 2.7.9   Multiple constraint synchronization.

the final operation. More than anything else, the cycle time through a serial flow process is governed by lot size and the ability of production operations to stay in sequence. As a general rule, setups must be sequence independent. Local machine optimization based on sequence-dependent setups will adversely affect global system performance based on mean flow time or minimum makespan criteria. For a high-mix, low-volume manufacturing environment, the master production schedule must be established by forward-scheduling

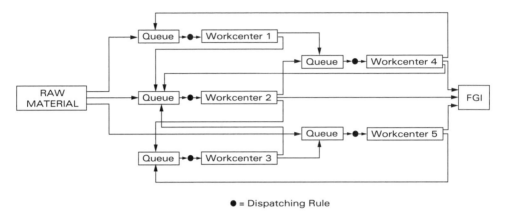

• = Dispatching Rule

Figure 2.7.10 Job shop.

constraint operations using sound scheduling methods. Due to the variability inherent in any manufacturing environment, changes to the planned schedule will be necessary and will result in a schedule that is suboptimal. Customer order cancellations, customer order expedite requests, stockouts, and difficult-to-predict process variability are the driving forces behind the need to modify an existing production schedule. While scheduling is predictive, change is reactive. Actions must be taken to resolve the deviation between the predicted and actual performance of the schedule. The flexibility to respond to change is a function of the following four criteria:

1. The placement of the master production schedule time fences
2. Automated flexibility to change schedules based on the difference between the earliest and latest start time using a combination of forward- and backward-scheduling techniques
3. Interactive manual scheduling flexibility to override the automated schedule
4. Flexibility of available capacity levels (e.g., overtime, subcontracting).

Events such as process defects and machine failure will reduce available capacity. Less than optimal quality and machine availability levels must be planned for when scheduling production operations. A capacity buffer (i.e., excess capacity) will reduce the likelihood that production schedules will have to be changed. Every effort must be made to achieve the *daily* production schedule. For job shop high-mix, low-volume manufacturing environments, the flow of work is not unidirectional. A particular machine may serve the roles of a gating operation and final assembly operation. An example of a job shop machine configuration is given in Fig. 2.7.10.

The number of possible sequences for any one particular machine is $n!$. If the sequences are entirely independent, there are $(n!)^m$ possible sequences for the n-job, m-machine job shop operations problem. The actual number of possible sequences is less than $(n!)^m$ because of less than maximum actual routings and sequencing limitations due to resource contention. Resource contention will occur when multiple jobs compete for the same resource, limiting the feasible number of sequences. The actual number of possible sequences is only of academic interest. The magnitude of the job shop problem defies analytical modeling and implicit computer enumeration techniques.

The development of feasible suboptimal schedules for a job shop is most often based on sequences that maximally reduce idle time at each workcenter. These schedules are referred to as a nondelay schedules. One process by which a nondelay schedule is created involves the use of a Gantt chart. During the process of creating a schedule, resource contentions (i.e., where two or more operations simultaneously occupy the same machine) are not permitted. After ordering jobs at each machine in operations sequence, the operation start dates are adjusted (for example, shifted to the left) until the earliest start time is arrived at without requiring a change in the operation sequence. This schedule is referred to as a semiactive schedule. A subsequent process is performed where global start-date changes are made independent of operation sequence in order to maximally compact the schedule. This schedule is referred to as an active schedule. The optimal schedule is a member of the set of all possible active schedules. Further reductions in idle time are possible through various job interchange and start time modifications to the Gantt chart. Once idle time is minimized, the resulting schedule is referred to as nondelay. Unfortunately, the set of all nondelay schedules may or may not contain the optimal schedule, and there is no way to know. The master production schedule is typically developed based on the nondelay schedule. Heuristic algorithms such as branch and bound and complex integer linear programming formulations have been developed to systematically create active and nondelay schedules.

On the production floor, job shop scheduling decisions are only made when a machine becomes idle. The only decision to be made is whether to produce a particular job in queue or allow the machine to remain idle. The decision to produce a particular job waiting in queue is based on dispatching rules and a schedule is unnecessary. Dispatching rules respond to dynamic behavior on the production floor, for instance, unplanned machine downtime, stockouts, or other factors that directly affect job status over time. Dispatching rules are used to develop a single local schedule for only one particular workcenter. Exhaustive research has been performed for dynamic job shop problems using computer simulation to measure the comparative performance of various dispatching rules. Some of the more important dispatching rules that have been developed are as follows, and this list is not exhaustive:

- **FCFS** (first come/first served): Produce jobs in the order that they arrived at the workcenter.
- **RND** (random): Each job has an equally likely chance to be produced.
- **SPT** (shortest processing time): The standard to which other dispatching rules are compared.
- **CR** (critical ratio): Produce the job with the smallest critical ratio based on the following calculation:

  CR = Time remaining until due date/total processing time remaining.
- **ODD** (earliest operation due date): Produce the job based on the following calculation:

  ODD = Number of operations/(final due date - gating process start time).
- **EDD** (earliest due date): Produce the job with the earliest final due date.
- **LWR** (least work remaining): Produce the job with the least total processing time remaining.
- **FOR** (fewest operations): Produce the job with the fewest work-center operations remaining.
- **HVF** (highest value job first): Produce the job that results in the highest total revenue.
- **NQ** (next queue): Produce the job at the operation with the smallest queue.
- **COVERT** (cost over time): Produce the job with the highest ratio based on the following calculation:

  COVERT = Total job cost/total processing time remaining.
- **S/OPN** (slack time per operation): Produce the job with the most slack time remaining based on the following calculation:

  S/OPN = Final due date - total processing time remaining.
- **MOD** (modified operation due date): Produce the job that is the maximum of the operation early completion time or the original operation due date.

There are problems associated with the use of computer simulations. A plethora of studies using a variety of assumptions with endless variations has often resulted in conflicting conclusions. Fortunately, factors such as the nature of arrival and service times and number of machines are not significant determinants when comparing the performance of various dispatching rules. What is important are the dispatching rules that have been widely reported to have desirable characteristics. There are several measures that are used to characterize the relative performance of various dispatching rules. Some of the more commonly used criteria for measuring dispatching rule performance are as follows:

- Profit maximization
- Dollar days late
- Maximum lateness
- Average lateness
- Lateness variance
- Mean flow time.

It is important to evaluate dispatching rules based on moderate to heavy loads. *Minimal management expertise is required to obtain good operations performance under conditions of light load.* A factory under duress from customers due to poor delivery performance may inappropriately create excess capacity through a combination of outsourcing, increasing the work force, and overtime, to improve delivery performance. This is not addressing the problem. This is a problem avoidance strategy that compresses gross profit margins and is indicative of incompetent management. There are numerous dispatching rules that can be effectively used to improve due date performance under conditions of moderate to heavy load. In general, performance to due date (i.e., maximum lateness, average lateness, lateness variance) is the most commonly used measure. Dispatching rules that consider due dates outperform FCFS and RND dispatching rules and are particularly effective if due dates are based on processing time. There is no conclusive evidence to support the notion that there is a statistically significant difference between dispatching rules based on due date performance criteria in an MRP environment. For MRP environments, ODD is the most logical choice and is the easiest to implement.

SPT is the best rule for achieving due dates under conditions of heavy congestion. Where time data is incomplete or nonexistent, due dates will have to be set based on alternative criteria. Under these conditions, SPT is remarkably insensitive to due date uncertainty as compared to other dispatching rules, in spite of its completely ignoring due dates. The penalty associated with SPT is the temporary excessive lateness that will occur for a few jobs. The SPT rule is without peer when performance is based on minimizing mean flow time and is distinctly superior for reducing the proportion of late jobs.

The use of dispatching rules in combination is superior to dispatching rules used alone if due dates are based on job processing times. Combining the superior mean oriented behavior of SPT with the superior variance oriented behavior of S/OPN yields a significant improvement in due date performance as compared to the use of either dispatching technique alone. The performance of the combination rule SPT and CR is not statistically significantly different than the combination rule, SPT and S/OPN. Combination rules require a weighting parameter that may be difficult to optimize for a particular, given situation.

Expediting will degrade all aggregate measures for all dispatching rules.

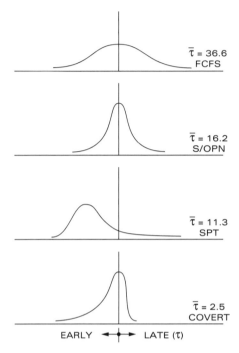

Figure 2.7.11 Lateness distributions.

Expediting is based on the notion that meeting a few customer-revised due dates is more important than achieving aggregate factory performance measures. Expediting a few jobs under conditions of heavy congestion will normally create additional emergencies and thus start a vicious cycle of expediting. This cycle can result in a situation where every job is expedited, and if allowed to continue unchecked, will result in a factory full of jobs with double or triple red-tag expedite priorities.

It is important to note that HVF is a dangerous priority rule to use if product cost is in direct proportion to labor content. Such jobs represent the longest processing time (LPT) and will significantly increase shop floor congestion. This will result in reduced throughput. COVERT attempts to ensure that the high-value jobs are completed on time and is superior to HVF when cost-accounting systems use direct labor as the product-costing driver. Cost-based priority rule performance measurements are often based on dollar days late. Dollar days late is the dollar value of the job multiplied by the days until the due date. For example, a $2,000 job that is five days late is equivalent to a $10,000 job that is one day late. Based on a simulation experiment performed by Carroll,[28] lateness distributions for several dispatching rules obtained from an eight machine pure job shop operating at 80 percent utilization are depicted in Fig. 2.7.11.

Two major assumptions used for job shop simulation studies are that machine setups are not sequence dependent, and due dates are set once and not subsequently revised. The latter assumption is contrary to MRP II practice. There is an abundance of priority rules from which a practitioner may choose. A determination must be made as to their appropriateness for a given situation. Priority rule behavior depends heavily on the manner in which due dates are set, level of shop load, and tightness of due dates. It will most likely never be possible to offer a single best policy to use for all cases.

When due dates are based on job processing times, the ability to tighten due dates and lead times is greatly enhanced. This results in lower work-in-process levels and an increase of available capacity. It is critical to reduce processing time variance. If the coefficient of variation (i.e., the ratio of the standard deviation of the processing time to the mean processing time) for processing times is one or greater, the production process is out of control and in need of major changes.

The purpose of all scheduling methods is to meet customer delivery date commitments at a profit. Late customer deliveries consume a significant amount of time and cost. Customers will have to be updated on the status of their order and the new delivery date. Costly expediting will suboptimize factory performance and adversely affect other orders yet to be produced. In some cases, an irate customer may be offered a discount whether or not this was contractually agreed to. Premium freight cost may also be incurred. There is a real potential to damage a company's reputation if delivery performance is chronically poor. This may result in the loss of customers. Adherence to schedule or dispatching rule priorities requires effective leadership and a well-trained and disciplined work force.

2.8 Test

There are many instances in which it is necessary to determine whether or not something or somebody meets specified conditions. For example, there are medical tests designed to determine whether a person has a particular disease and tests on electronic products to decide whether the product will perform within specified conditions. In any test situation, one wants a high probability of detecting a faulty device (or disease) and, simultaneously, a high probability of passing a fault-free device (or healthy person). Because of inescapable variation, two types of diagnostic errors are made. The first is the false alarm. False alarms occur when something that is fault-free is found faulty by the test method. The second is the missed fault. Missed faults occur when something that is truly faulty is not detected by the test method. Depending on the situation, both types of errors can be disastrous. In general, false alarms cause effort to be expended in repairing a problem that doesn't exist and reduce

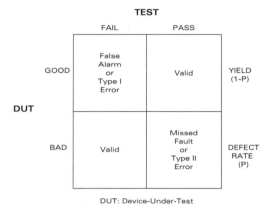

Figure 2.8.1 The two types of diagnostic errors.

available capacity. False alarms may also result in a missed customer delivery date. Missed faults, on the other hand, are often not discovered until some time after the testing process. In the case of an electronic product, the fault may be found by the customer. This can result in great expense or injury and the loss of customer loyalty. To be competitive, high diagnostic error rates cannot be tolerated.

Test effectiveness, the risk and cost consequences of diagnostic errors, must be evaluated prior to considering alternative test strategies. The general test effectiveness problem will be discussed in the context of electronic devices, but as previously noted, the method applies to any test situation. The chart of Fig. 2.8.1 shows four possible situations that can occur when a device is tested.

There is a true defect rate, P, for the device under test. A desirable test method will pass truly good devices and fail truly bad ones. The two diagnostic errors occur when the test fails a good device (false alarm) or passes a

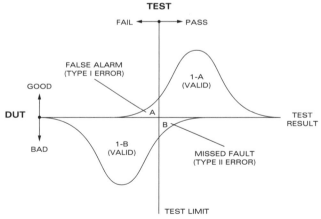

Figure 2.8.2 The two types of diagnostic errors with variation in the result.

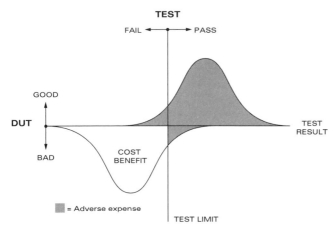

Figure 2.8.3 The economics of parametric test.

bad device (missed fault). These are directly analogous to the Type I and Type II errors that are made in statistical hypothesis tests. Fig. 2.8.2 expands on the four test situations by showing that there is uncertainty associated with the test results.

Think of Fig. 2.8.2 as applying to a single parameter being tested on a single device. The X-axis is the test result on a continuous measurement scale. The Y-axis is located at a test limit. There could be two test limits, but for simplicity only one test limit is shown. The four quadrants in Fig. 2.8.2 correspond to the four situations shown in Fig. 2.8.1. The distributions reflect that there is variation in either the test method (i.e., measurement error) or in the response of the device under test, or both. Other sources of variation can be included such as device-to-device and test system variation. The shape of the distribution is not important. What is important is the false alarm rate (the probability, A) and the missed fault rate (the probability, B). That there are error distributions for both good and bad devices is shown by distributions both above and below the X-axis. Whether these distributions are the same, as depicted in Fig. 2.8.2, or different is unimportant. It is assumed that the variation obtained in repeatedly testing the same parameter is random. However, it is not necessary to assume that the variation is stable and predictable across several devices; although this is certainly desirable from a process control standpoint.

Fig. 2.8.3 further expands on the four test situations by showing the cost benefits and undesirable expenses incurred at subassembly test for each discrete parametric test. Clearly, if a truly faulty device is detected, there is a cost benefit. The testing of truly good devices and diagnostic errors are undesirable expenses that are incurred using any manufacturing test strategy. One way to quantify the overall test effectiveness of independent tests is shown in Fig. 2.8.4.

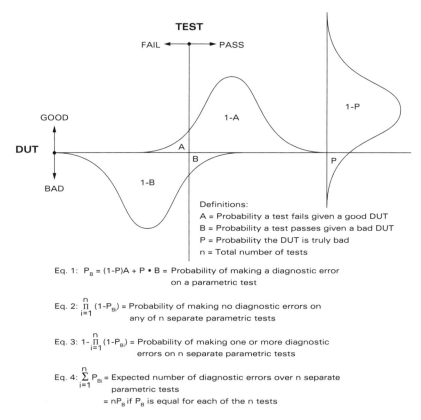

TEST

FAIL ◄———•———► PASS

GOOD

1-P

1-A

DUT

A

B

P

BAD

1-B

Definitions:
A = Probability a test fails given a good DUT
B = Probability a test passes given a bad DUT
P = Probability the DUT is truly bad
n = Total number of tests

Eq. 1: $P_B = (1-P)A + P \cdot B$ = Probability of making a diagnostic error
on a parametric test

Eq. 2: $\prod_{i=1}^{n} (1-P_{Bi})$ = Probability of making no diagnostic errors on
any of n separate parametric tests

Eq. 3: $1 - \prod_{i=1}^{n} (1-P_{Bi})$ = Probability of making one or more diagnostic
errors on n separate parametric tests

Eq. 4: $\sum_{i=1}^{n} P_{Bi}$ = Expected number of diagnostic errors over n separate
parametric tests
= nP_B if P_B is equal for each of the n tests

Figure 2.8.4 Quantification of overall test effectiveness.

These tests might be different parametric tests on the same device or tests of the same parameter on different devices. Using the test of an electronic printed circuit board as an example, the distribution aligned vertically in the diagram represents the distribution of boards for a particular parameter. Thus, P is the proportion of boards that are faulty with respect to a certain parameter. P_B is the weighted average of the two diagnostic errors A and B, and would be the probability of making a diagnostic error of either type on a particular parametric test. As an example, suppose P=0.2, A=0.05, and B=0.10. Then P_B=0.06. The expected number of errors on this particular test for 1,000 printed circuit boards would be 60 (Eq. 4). Equations 2 and 3 are useful when thinking about diagnostic errors across different parametric tests on the same printed circuit board. In this case, P, A, and B would likely differ from one test to the next, so that P_B would vary from test to test. Note that equations two through four are only valid if the *n* test results are independent. Depending on the parametric tests involved, it may be necessary to modify the test sequence based on range parameters among adjacent tests to achieve independence. The equations in Fig. 2.8.4 may be used to quantify

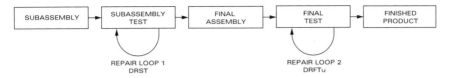

Figure 2.8.5 The classical manufacturing test strategy.

test effectiveness in terms of probabilities or the expected number of errors.

Many high-mix, low-volume manufacturers face the problem of insufficient capacity for their test process. When the present strategy is to perform subassembly test prior to final test, one alternative is to final-test the product first and then send only known defective subassemblies to the subassembly test process, thus increasing the capacity of the subassembly test process.

Fig. 2.8.5 represents the production flow of subassemblies used by many high-mix, low-volume production environments. Assembled subassemblies flow directly to subassembly test after which they are final assembled into the finished product and then passed to the final test process. Over time, many manufacturers reach capacity at the subassembly test level. This bottleneck occurs due to growth in subassembly volume. Increased volume occurs when sales increase or new subassemblies are added to the existing product line. Such a scenario is commonplace for a manufacturer pursuing a competitive advantage in agility. Generally, a two-prong approach has been taken to increase test capacity—improve throughput through the existing test process and/or purchase additional test capacity. In many cases, the purchase of additional capacity at subassembly test is extremely expensive. This is particularly true for electronic equipment manufacturers.

An alternative test strategy is to bypass the subassembly test completely and final-test the finished product first. Consider Fig. 2.8.6. All good products are subsequently shipped directly to the customer, while only the defective subassemblies are routed to the subassembly test process. Subassembly test takes

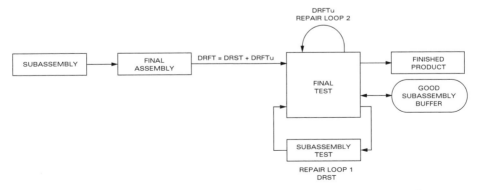

Figure 2.8.6 Manufacturing test strategy for increasing test capacity.

on the troubleshooter role. This test strategy, coupled with high yields at the final test level, can significantly reduce the volume of subassemblies passing through subassembly test and increase the overall capacity of the test process.

An analysis can be performed such that the cost to implement the test process of Fig. 2.8.6 can be expressed in terms of the cost for the test process of Fig. 2.8.5. This analysis can help to determine the best decision for reducing the bottleneck at subassembly test. Consider the test process of Fig. 2.8.5. The cost at subassembly test can be defined as:

$$ST\$ = Cgst\ (1\text{-}DRST) + Cbst\ (DRST) \qquad (2.8.1)$$

Where: $ST\$$ = Total cost per subassembly at subassembly test
  $Cgst$ = Total cost to test a good subassembly at subassembly test
  $Cbst$ = Total cost to test and repair a bad subassembly at sub-assembly test
  $DRST$ = Defect rate at subassembly test (fractional).

The total cost at final test can be defined as:

$$FT\$ = Cgft\ (1\text{-}DRFT_u) + Cbft\ (DRFT_u) \qquad (2.8.2)$$

Where: $FT\$$ = Total cost per board at final test
  $Cgft$ = Total cost to test a good board at final test
  $Cbft$ = Total cost to test and repair a bad board at final test
  $DRFT_u$ = Defect rate at final test (fractional).

The total cost for the test process of Fig. 2.8.5 is therefore:

$$TOTAL\$ = Cgst\ (1\text{-}DRST) + Cbst\ (DRST) + Cgft\ (1\text{-}DRFT_u) + Cbft\ (DRFT_u)$$
$$(2.8.3)$$

Consider the test process of Fig. 2.8.6. All subassemblies pass directly to the final test level. The defects caught at subassembly test *(DRST)* from Fig. 2.8.5 will now be passed to the final test level. The defect rate at final test is now the sum of *DRST* and *DRFT_u,* provided that all defects occurring at subassembly test can be discovered at final test. All defects caught at the final test level *(DRST + DRFT_u)* will be isolated to the subassembly level and routed to subassembly test. The defective subassemblies sent from final test to subassembly test will then be replaced from a buffer of known good subassemblies at final test. In this manner, the production plan and customer due dates will be achieved with minimal interruption. All defects are now at subassembly test.

Subassembly test will be able to repair the *DRST* defects; however, the $DRFT_u$ failures will not be detected at subassembly test and will have to be repaired at the final test level (repair loop two). The reason for this is that $DRFT_u$ failures manifest themselves due to less than 100 percent fault coverage at subassembly test. Subassembly test detects most process faults but interactive and dynamic faults at the finished product level may not be detected. After all defective subassemblies are repaired and subsequently verified as good at final test they are returned to the good subassembly buffer. The total cost at subassembly test can be defined as:

$$ST\$ = Cgst \ (DRFT_u) + Cbst \ (DRST) \tag{2.8.4}$$

A new cost, *Cx*, has to be added to final test. *Cx* represents the cost to test and isolate a failure at the subassembly level, remove and replace the defective subassembly, and retest the repaired subassembly. The total cost at final test can be defined as:

$$FT\$ = Cgft \ (1\text{-}DRST\text{-}DRFT_u) + Cbft \ (DRFT_u) + Cx \ (DRST+DRFT_u) \tag{2.8.5}$$

The total cost for the test process of Fig. 2.8.6 is therefore:

$$TOTAL\$ = Cgst \ (DRFT_u) + Cbst \ (DRST) + Cgft \ (1\text{-}DRST\text{-}DRFT_u) + Cbft \ (DRFT_u)$$
$$+ \ Cx(DRST+DRFT_u) \tag{2.8.6}$$

Costs and defect rates are the variables that determine which test process, Fig. 2.8.5 or Fig. 2.8.6, is more cost-effective. When the cost to implement the process of Fig. 2.8.5 equals the cost of Fig. 2.8.6, the test process of Fig. 2.8.6 is preferable, because the purchase of additional capital equipment is not required. If the costs are fixed and the defect rates at subassembly test and final test are varied for Eqs. (2.8.3 and 2.8.6), the relationship between defect rate at subassembly test and the defect rate at final test can be derived when the test process of Fig. 2.8.6 is less than or equal to the cost for the test process of Fig. 2.8.5.

The incremental cost to isolate and repair a defect at final test as opposed to subassembly test is approximately a factor of ten. This is not always the case. The incremental cost may be as low as a factor of four or lower. There are several factors that affect the incremental cost to repair a defect at final test as opposed to repairing it at subassembly test:

- Subassembly complexity
- The level of subassembly self-diagnostic capability

- The level of test process fault isolation capability
- The wage and overhead differentials at final test versus subassembly test
- The level of mature technical talent at the final test level.

To understand the conditions under which the process of Fig. 2.8.6 is preferable, the difference between Eq. (2.8.3) (the cost for the test process of Fig. 2.8.5) and Eq. (2.8.6) (the cost for the test process of Fig. 2.8.6) can be calculated algebraically:

$$
\begin{aligned}
\textit{If:}\quad Cgst &= x \\
Cbst &= f_1\,(Cgst) = f_1\,(x) \\
Cgft &= f_2\,(Cgst) = f_2\,(x) \\
Cbft &= f_2\,(Cbst) = f_1\,(f_2)(x) \\
Cx &= 2(Cgft)+R = 2(f_2)(x)+R
\end{aligned}
$$

Where: x = Cost of testing a good subassembly at subassembly test
  f_1 = The incremental cost factor (ICF) for bad versus good within either subassembly or final test
  f_2 = ICF for final test versus subassembly test
  R = Removal and replacement cost at final test.

Then:

$$
\Delta = (eq.\ 2.8.3 - eq.\ 2.8.6) = x - [x + f_2(x) + R]DRST - [x + 2(f_2)(x) + R]DRFT_u
$$

$$(2.8.7)$$

If $\Delta > 0$ then the test process of Fig. 2.8.6 will cost less than the test process of Fig. 2.8.5. Note that Δ does not depend on f_1, the ICF for bad $(C_b..)$ versus good $(C_g..)$ within either subassembly test or final test given the assumption of equal f_1's (a sound assumption based on case studies).

The relationship between $DRST$ and $DRFT_u$ when $\Delta = 0$ is found in Fig. 2.8.7. ICF values of 4 and 10 are used for f_2, $x = \$5$, and $R = \$10$. The test process of Fig. 2.8.6 has lower cost when the defect rates lay below the associated line for each ICF. Now that the threshold relationship between the defect rate at subassembly test $(DRST)$ and the defect rate at final test $(DRFT_u)$ has been identified, we can use these results to consider the process of Fig. 2.8.6 more carefully.

Fig. 2.8.8 represents the relationship (for the test process of Fig. 2.8.6) between the percentage turn-on rate at final test, $100(1-DRST-DRFT_u)$, and the percentage fault coverage at subassembly test, $100(DRST)/(DRST+DRFT_u)$, for $\Delta = 0$ at ICF values of 10:1 and 4:1. For an ICF of 4:1, the turn-on rate at final test may be as low as 91 percent and the test process of

f_2	R	x	DRST	$DRFT_u$	% Fault Coverage at Subassembly Test 100 [DRST/(DRST + $DRFT_u$)]	% Turn-on Rate at Final Test for Fig. 2.8.6 100 (1 - DRST - $DRFT_u$)
10:1	10	5	0	.04	0	96
10:1	10	5	.08	0	100	92
4:1	10	5	0	.09	0	91
4:1	10	5	.14	0	100	86

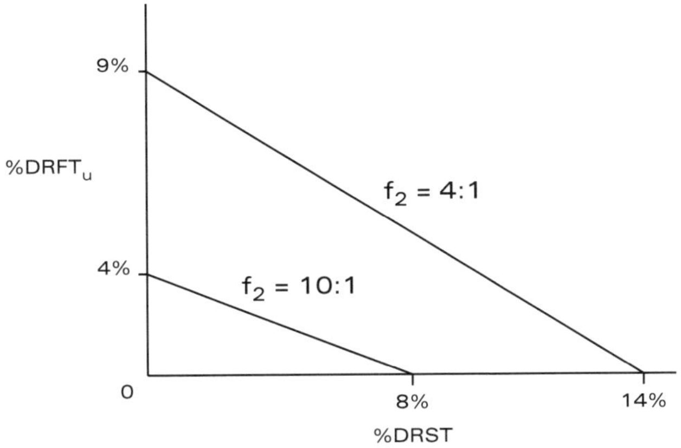

Figure 2.8.7 The conditions that must exist for the manufacturing test strategy of Fig. 2.8.6 to be economical.

Fig. 2.8.6 is justified, even if none of the defective subassemblies sent to subassembly test will be repaired (0 percent fault coverage). Clearly, if 100 percent of the defective subassemblies are repaired at subassembly test, an even lower turn-on rate can be tolerated at final test (86 percent). For an ICF (final test versus subassembly test) of 10:1, a higher turn-on rate is demanded at the final test level. Fault coverage for subassembly test is typically in the range of 85 to 99 percent (shaded portion). Thus, we can now see the relationship between the percentage turn-on rate at final test and the percent fault coverage at subassembly test that must exist for the test process of Fig. 2.8.6 to be preferable. The global turn-on rate for all subassemblies in a high-mix, low-volume manufacturing operation is seldom as high as 100 percent; however, certain subassemblies may exist that do exhibit at least 85 percent turn-on rates. These particular subassemblies can be selected for the test process of Fig. 2.8.6 when the ICF is acceptably low.

Knowledge of the relevance of process control and an understanding of the ingredients and dynamics of the defect cost spectrum are necessary when analyzing alternative test strategies. While it is highly desirable to have statistical control across all processes in a high-mix, low-volume manufacturing operation, it is not a necessity for the successful implementation of the test strategy of Fig. 2.8.6. However, if a process is not in statistical control, defect rates are not predictable and conversion to the test process of Fig. 2.8.6 may

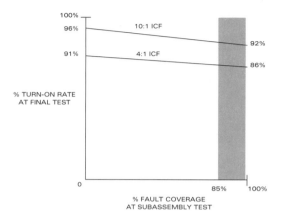

Figure 2.8.8 The conditions that must exist for the manufacturing
test strategy of Fig. 2.8.6 to be economical.

not be cost-effective in the future. A complete characterization of the defect cost spectrum should be performed on a subassembly by subassembly basis, targeting the population of subassemblies with the highest cumulative test time. Cumulative test time is the tactical weapon that should be used to obtain the greatest leverage for increasing test capacity and thus improving on-time customer delivery performance.

Fig. 2.8.9 represents how design, material, and process defects contribute to overall manufacturing defect cost. The product defect dollar percentages are based on an actual case study. There must be very few actual design problems to economically produce a product. The design and material interaction

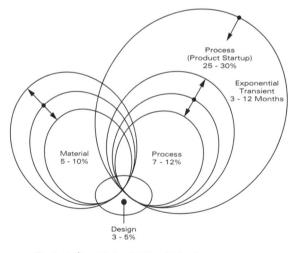

Defect $ = 15 to 25% of Ideal cost
Actual cost = Ideal cost + defect $

Figure 2.8.9 Defect cost spectrum.

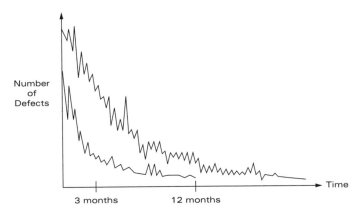

Figure 2.8.10 Process faults versus time.

problems are typically reduced very little during the lifetime of a product, and relatively few design and process problems occur. The extent of the material problems is determined during the design of the product. Only the size of the fluctuation in material defects will change. Incoming quality levels from various vendors affect the magnitude of the material defect fluctuations.

Consider Fig. 2.8.10. When a new product is introduced into manufacturing an initial process defect transient occurs and then a steady state is exponentially approached. During the steady-state condition, only the size of the fluctuation in process defects will vary. The magnitude of the transient is a function of product complexity and can be affected by proper process design for the incoming product. Process control can only affect the steady-state condition. New subassemblies should not be considered for possible implementation of the test process of Fig. 2.8.6. Only subassemblies that have reached the steady-state condition should be considered. The worldwide trend to reduce defects and defect repair costs using statistical quality control techniques will improve product quality, reduce manufacturing production cost, increase available capacity, increase throughput, and improve on-time delivery performance.

CHAPTER 3

Cross-Functional Relationships

3.1 Customer Order Dimensions

In order to significantly reduce the uncertainty of demand for any manufacturing environment, it is critical to understand the underlying intrinsic and extrinsic causes of demand variability. Intrinsic demand variability is not controlled by the company and represents the true underlying demand. Intrinsic demand variability can be stratified based on the product, geography, customer, market, and competitive environment (e.g., monopolistic, oligopolistic, competitive). Such an analysis should be performed for at least a one-year period in order to properly evaluate intrinsic factors such as seasonality, economic trends, competitor performance, and customer linkages.

Extrinsic demand is directly controlled by the company and can have a significant effect on overall demand variability. Extrinsic demand variability is associated with marketing, sales, and manufacturing. Marketing and sales are primarily focused on increasing volume, market share, profit, and customer satisfaction levels. Marketing and sales induce extrinsic variability through discounting structures, sales promotions, and sales quotas. The resulting demand pattern may exhibit sharp peaks and severely limit the company's ability to respond. Discounting structures and sales promotions seldom result in incremental profits and should rarely be employed. Changes in price for particular products will affect demand differently. Consider Eq. (3.1.1):

$$Elasticity = \frac{percentage\ change\ in\ quantity\ demanded}{percentage\ change\ in\ price} \qquad (3.1.1)$$

Elasticity measures are used to evaluate the change in revenue that occurs due to a change in price. If the result of Eq. (3.1.1) is one, the product has unit elasticity. If the result is greater than one, the demand is elastic and if the result is less than one, the demand is inelastic. A product's sensitivity to a price change is not always known. In this case, estimates of total sales revenue at a given price determine the demand curve and its demand elasticity. The sales estimate is also used to determine the break-even point based on the total investment made for a newly introduced product. Once the impact of a price change for a particular product is known, this information can be used to predict the outcome of a competitor's price change. If the demand is inelastic the company may choose not to respond in kind. Knowledge of a product's demand elasticity can offer a competitive advantage over less knowledgeable competitors. If discounting for a large customer order or sales promotion is going to occur, manufacturing must be notified in advance. Short-term peaks in sales must be planned for and incorporated in the master production schedule. A significant overload may require the use of overtime or outsourcing.

The time periods for which the achievement of sales quotas are measured will induce significant variability into the underlying intrinsic demand pattern. Demand will peak at the end of the time period for which sales quota performance is measured. This results in what is referred to as a "hockey stick" demand pattern. To effectively reduce demand volatility associated with sales quotas, the sales force should be rewarded for leveling factory orders and penalized for inducing variability. Manufacturing can be a source of variability if lead times are long. Long lead-time manufacturing environments are characterized by large lot-size production methods that inflate lead time, reduce responsiveness, and increase inventory levels.

For high-mix, low-volume manufacturing environments, the product portfolio must be evaluated based on revenue, profit contribution, and support costs, as compared to the business objectives of the company. This analysis should be performed on a quarterly or semiannual basis.

Consider Fig. 3.1.1. While support costs are typically larger for the low-revenue, high-mix products, standard cost accounting methods will allocate a disproportionate amount of support costs to the high-revenue, low-mix products. Although the profit margin for low-revenue, high-mix products is larger than the high-revenue, low-mix products, the profit contribution for low-revenue products will be overstated. This is misleading. A further analysis of the products showing a loss (products C, F, G, H, L, M, N, and O of Fig. 3.1.1) is necessary in order to evaluate what action, if any, must be taken.

Consider Fig. 3.1.2. The introduction phase of a new product involves high marketing costs and high net losses. Failure rates may be high, and a period of time is required for the product to gain market acceptance. The losses for products C, F, and O (Fig. 3.1.1) are expected and corrective action is not

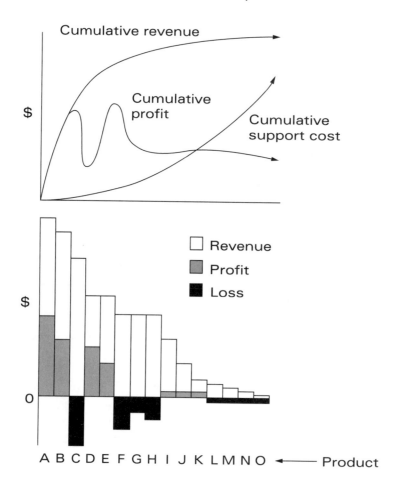

Figure 3.1.1 Product revenue, profit, and cost.

called for at this time. The growth period is characterized by increasing sales and profits begin to rise. Profits may diminish if competitors are also entering the marketplace during this period. The time-to-market for a new or enhanced product is critical in this regard. Product quality, features, and brand are emphasized to sustain growth. The failure of products H and G to show a profit during the growth phase is a concern and should be monitored closely. During the initial stage of the maturity phase, sales continue to increase, but at a slower rate than the growth phase. Competition is intense and marginal competitors usually drop out of the market. During the battle for customers, profits will tend to get "squeezed" as price reductions and increased advertising and promotion efforts are pursued to maintain marketshare.

A cost reduction effort may be invested in for a mature product in order to

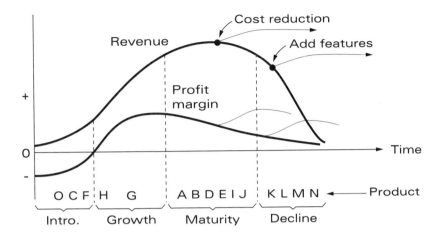

Figure 3.1.2 Product life cycle.

increase profits. Cost reduction efforts usually focus on reducing material cost. Products A, B, D, E, I, and J are profitable and a cost reduction effort may make sense for products E and I based on their respective profit positions. Product J is too close to the decline phase to allow for sufficient time to recoup the cost reduction investment. During the decline phase, sales and profits decline and serious consideration is given to whether a particular product should be obsoleted. Competitors who remain during the decline phase usually make only a small profit or break even. Products K, L, M, and N are not contributing adequate profit but are consuming valuable resources and should be phased out. A decision to invest in the enhancement of products E and I should be seriously considered. Such an investment will significantly extend the maturity phase for these particular products at a lower cost than a new product design requires. Business management must establish a strategy for production and marketing investment for a product in a shrinking marketplace. Based on revenue and profit results, the phasing out of a particular product may be the only option and is a decision that must be made. The pruning of products that are losers is a tactic used by corporate turnaround artists to get losing companies back on track. For high-mix, low-volume manufacturing environments in the electronics and steel industries, it is not uncommon for product life cycles to extend 10 years or more.

The cross-functional relationships among marketing, sales, and manufacturing must be more than simply an interface. Marketing, sales, and manufacturing must be cross-functionally integrated in order to effectively control and respond to the extrinsic dimension of customer order demand variability.

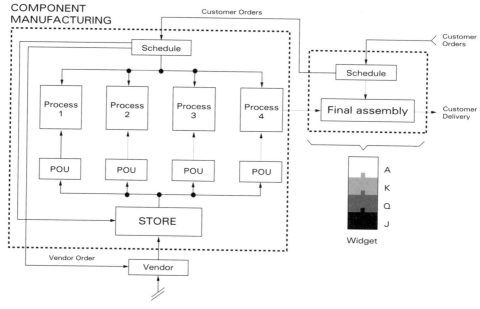

Figure 3.2.1 Value delivery chain.

3.2 Lot Sizing Simulation

The main purpose of manufacturing simulation is to gain an increased under-standing of the relationships among diverse variables in the course of achiev-ing some measure of effectiveness. Simulation starts with a simplified model of a real situation, thus allowing insights to be gained through experimenting with a variety of inputs. Although the major focus of simulation has been on stochastic system behavior, the following simulations will be deterministic (i.e., fixed outcome). Simulation enables the individual to become more involved in the study, thereby increasing the liklihood of acceptance of the study's assumptions and conclusions. The following simulations demonstrate a variety of manufacturing system behaviors and describe the necessary ele-ments that must be considered for making effective manufacturing operations decisions. Note that the nomenclature used to identify products for the fol-lowing simulations is J, Q, K, A. These symbols are directly analogous to common playing cards (i.e., Jack, Queen, King, Ace) which can be used to physically perform the simulations presented in the text.

A partial explosion (i.e., limited detail) of a conceptual value delivery chain is given in Fig. 3.2.1. The flow of information and the characteristics of this value delivery chain are as follows. A customer places orders for widgets from a build-to-order final assembly operation. Once orders are received, final assembly will develop an assembly schedule, promise a delivery date to the

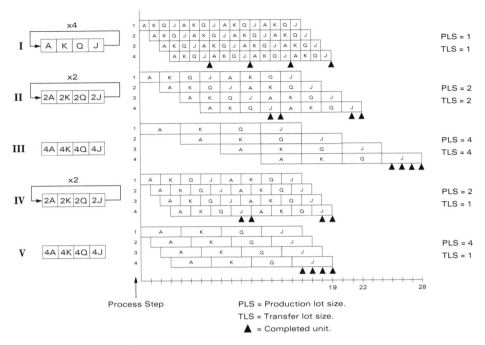

Figure 3.2.2 Lot sizing.

customer, and place an order for components to its component manufacturer. The component manufacturer is a build-to-order manufacturing environment. The component manufacturer will develop a production schedule, promise a delivery date to final assembly, and place an order to its vendor for raw materials. The vendor is a build-to-stock manufacturer that supplies raw materials to the component manufacturer's store. The store supplies raw materials to each production process point-of-use (POU) stock location based on the manufacturing production schedule. The final assembly and component manufacturing operations are initially devoid of inventory.

Let us now consider the case where an order for four widgets is placed at final assembly, and each widget consists of one each J, Q, K, and A component parts. The total demand placed on the component manufacturer is four each of J, Q, K, and A component parts, and the processing time for each component part at each process step is equal to one time unit. A total of four process steps is required to produce each component part. There are many ways in which the component manufacturer can choose to respond to this demand, and each choice will not produce equally significant results.

Five possible schedules are given in Fig. 3.2.2. Schedule I produces to the sequence A, K, Q, J and iterates this schedule four times. The production lot size for each component part is one. The production lot size is that quantity of a particular item that is produced before a dissimilar item is produced.

Schedules II and IV produce to the sequence 2A, 2K, 2Q, 2J and iterate this schedule two times. The production lot size for each component part is two. Schedules III and V produce to the sequence 4A, 4K, 4Q, 4J that is only performed once. The production lot size for each component part is four. Schedules I, II, and III of Fig. 3.2.2 have transfer lot sizes equal to their respective production lot sizes. The transfer lot size is the quantity of component parts required before a transference to the next process step is allowed (i.e., the previous process step will have to completely finish a production lot prior to sending it to the next process step). The Gantt charts are developed for each production schedule, and it is clear that schedule I is superior in terms of responsiveness. Schedule I will begin initial customer deliveries sooner than all other scheduling choices. Responsiveness is a measure of the time required to negotiate (i.e., produce) and deliver at least one unit of the entire mix of products ordered.

The responsiveness performance of schedules II and III can be significantly improved by setting their respective transfer lot sizes equal to one. The resulting makespans are now equal to the makespan of schedule I, and their resulting Gantt charts are shown in Fig. 3.2.2 (schedules IV and V). Although makespan and responsiveness are dramatically improved over schedules II and III, the resulting schedules (IV and V) are not as responsive as schedule I. Production lot sizes greater than one are often required to amortize the adverse effects of setup time. In many cases, the failure to amortize setups will result in an inordinately long makespan. It is critical to have transfer lot sizes that are less than their respective production lot sizes due to the improvements in makespan and responsiveness performance that occur. If the time required to transfer a lot from one process step to the next is excessive, the benefit of small transfer lot sizes is negated. It is essential to have operations process steps that are in close physical proximity to their respective feeder process steps in order to realize the benefits of small transfer lot sizes.

The inventory position and responsiveness of the component manufacturer are directly related to the lot sizing rules employed. Consider the case where the final assembly and component manufacturing operations are initially devoid of inventory, order arrivals are inexhaustible, and all schedules are iterated ad infinitum. The inventory position of the component manufacturer is shown for each schedule (Fig. 3.2.2) in Fig. 3.2.3. The best possible inventory position results from the use of schedule I. For each doubling of the production lot size, the inventory level will also double, under the condition that production lot sizes (PLS) are equal to the transfer lot sizes (TLS). If the transfer lot size is less than the production lot size, the resulting inventory increase is less in terms of maximum inventory level and absolute deviation of the inventory level as compared to the case of equal production and transfer lot sizes.

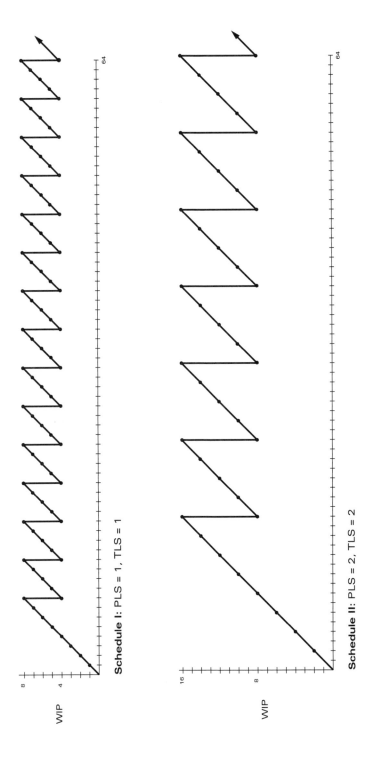

Figure 3.2.3 Inventory for Fig. 3.2.2 with inexhaustible arrivals.

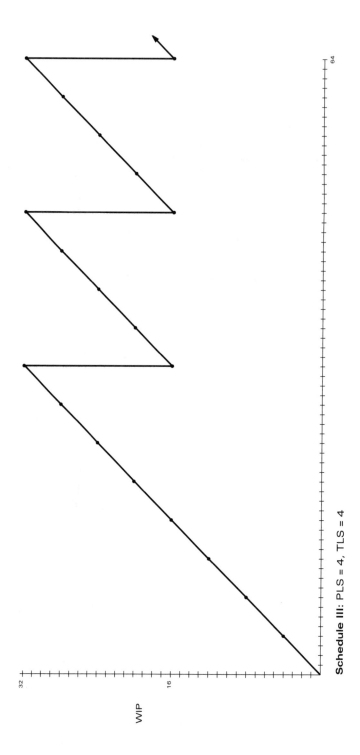

Schedule III: PLS = 4, TLS = 4

Figure 3.2.3 (*continued*) Inventory for Fig. 3.2.2 with inexhaustible arrivals.

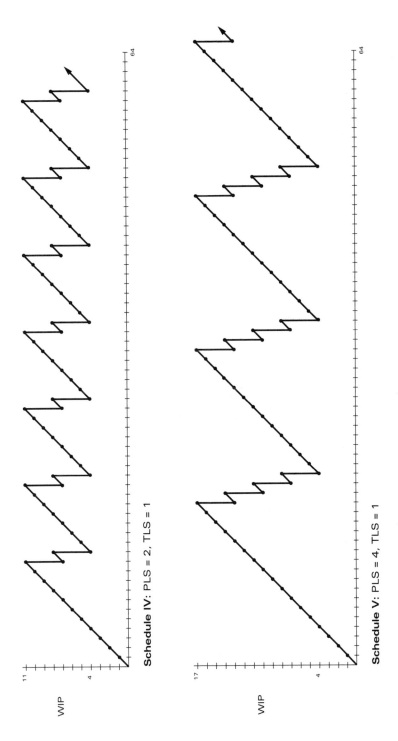

Figure 3.2.3 *(continued)* Inventory for Fig. 3.2.2 with inexhaustible arrivals.

In order to correctly determine manufacturing lead time, it is necessary to establish the steady-state condition for each schedule as a starting point. The steady-state condition is achieved at the first point in time that each manufacturing process step is being utilized. For schedules I through V the steady-state condition is reached at time units 4, 7, 13, 4, and 4 respectively and the manufacturing lead times are 4, 8, 16, 7, and 13 respectively. When the production lot size is equal to the transfer lot size (schedules I, II, III), work-in-process inventory and lead time are the same thing (i.e., maximum WIP = lead time). Lead time and responsiveness are inversely related. Decreased lead times result in a consequent increase in responsiveness performance and vice versa.

It is essential to realize that the preferable schedule for the component manufacturer is not exactly the same as the customer's order sequence. Consider the case where a customer orders 20A, 20K, 20Q, and 20J, and the component manufacturer produces to this sequence. The inventory position and responsiveness performance of the component manufacturer will be severely degraded as compared to schedule I. Over time, incoming customer order quantities are likely to randomly change, and the resulting lot sizes will also randomly change. Random lot sizing will result in random inventory requirements and random responsiveness performance. Plainly and simply, attempting to respond to demand in the exact sequence that it is ordered is the epitome of incompetence and a prescription for disaster that must never be allowed to occur. Efficient and effective performance is only achieved by decoupling the component manufacturer from the customer through appropriate scheduling and lot sizing techniques.

Cross-functional linkages must be formed among all component manufacturer, final assembly, and raw material vendors in order to exploit scheduling strategies such as schedule I of Fig. 3.2.2. Inventory will be stockpiled at final assembly if final assembly promises customer delivery dates for periods earlier than the periods in which components will arrive. This problem is indicative of a component manufacturer and final assembly operation that are strategically isolated from each other. Component manufacturers and their respective customers must be strategically linked in order to optimize performance throughout the value delivery chain and prevent such problems from ever occurring. Raw material vendors must be strategically linked with their customers in order to facilitate the fast response and delivery times required. Linkages are most effectively established through electronic data interchange (EDI) methods of communication over geographic distances.

3.3 Inventory

Inventory is one of the most widely discussed and analyzed topics in business practice today. Investment in inventory is the single, largest investment a company makes, and from a financial standpoint, it has a dramatic effect on cash flow. To understand the strategic financial implications of inventory on the financial performance of a company, we can analyze a company's return on investment (ROI) based on the DuPont ROI model. The DuPont ROI model is based on the fundamental relationship shown by Eq. (3.3.1):

$$ROI = \textit{Profit percentage} \times \textit{Investment turnover} \qquad (3.3.1)$$

The DuPont ROI model is shown in Fig. 3.3.1. The required values are obtained from a company's balance sheet and income statement. Although inventory is found in the balance sheet and income statement branches of the model, investment turnover has a minimal effect on taxation and a significant effect on ROI. In a competitive environment of shrinking profit margins, profitability can be maintained by increasing the investment turnover rate. For illustrative purposes, arbitrary values have been added to the model which result in a 10 percent ROI. The finance department of any company can provide the necessary values to calculate their particular ROI.

Fig. 3.3.2. shows that a 10 percent reduction in inventory results in a 12 percent improvement in ROI. Although cost savings associated with this inventory reduction could be realized, the approach taken in Fig. 3.3.2 is conservative and holds these costs constant. The inventory sensitivity factor is 1.2:1. A 1 percent change in inventory results in a 1.2 percent increase in ROI. The sensitivity factor will vary based on the financial position of a particular company. An alternative to increasing ROI via investment turnover would be to increase sales.

A sales increase of 39 percent, along with a proportional change in relevant investments and costs, would be required to achieve the same ROI as a 10 percent reduction in inventory and is shown in Fig. 3.3.3. As a general rule, reductions in inventory will provide much greater leverage than comparable sales increases for improved ROI performance. Clearly, a 39 percent increase in sales volume would require significant capital investment to increase capacity unless, of course, a company is operating grossly under capacity. An unlikely method for obtaining a 39 percent increase in sales is to reduce prices and hope that the competition does not respond in kind. A modest inventory reduction of 10 percent can be achieved much more quickly and at a lower cost. A 10 percent reduction in inventory will most certainly draw much less attention from the competition than the price reductions required to increase sales by 39 percent. Management must not rely on short term sales performance

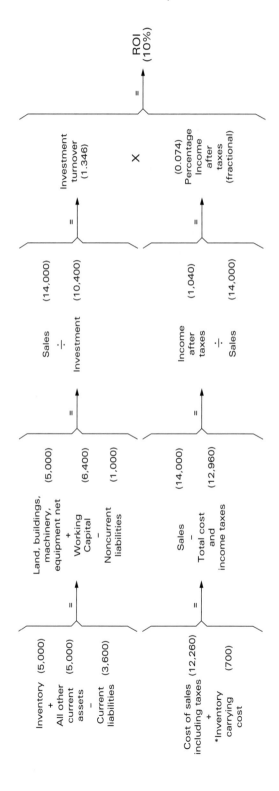

Figure 3.3.1 DuPont return-on-investment model (reference).

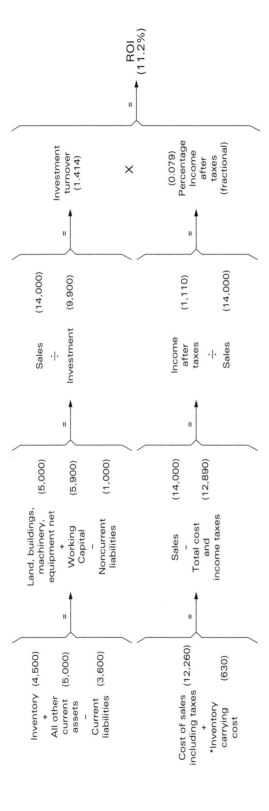

Figure 3.3.2 DuPont return-on-investment model (10 percent inventory reduction).

* Based on 14% weighted average cost of capital

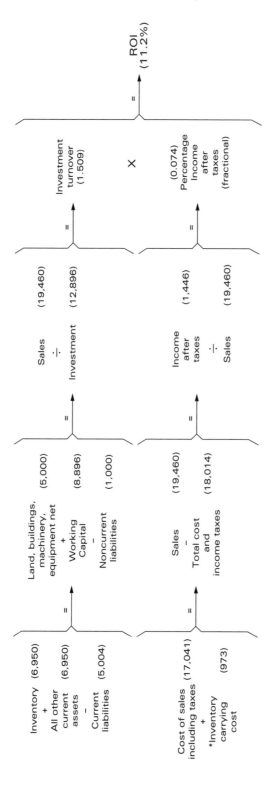

Figure 3.3.3 DuPont return-on-investment model (39 percent sales increase).

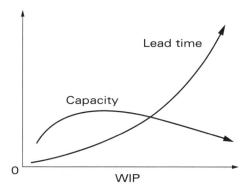

Figure 3.3.4 The relationships among WIP, lead time, and capacity.

(e.g., monthly) as the mechanism for improving ROI nor use sales as an excuse for poor inventory turnover performance. Management must focus on *reducing* the aggregate level of inventory investment.

Production planning and detailed scheduling are necessary to ensure that the *timing* of operations will satisfy customer demand while achieving the strategic objectives of the firm. Products are manufactured through multiple operations and the length of time required to complete all operations results in work-in-process (WIP) inventory. The length of time that WIP inventories spend in the manufacturing system is directly related to the lot size rule employed. WIP inventories are adverse expenses in terms of the capital investment, space requirements, and material handling costs incurred. In order to understand how investment in inventory can be lowered while maintaining current shipment levels, the drivers of inventory investment must be understood. The average total inventory ($\overline{I}$) is controlled by average capacity ($\overline{C}$) and the average manufacturing lead time ($\overline{L}$), as shown in Fig. 3.3.4 and expressed mathematically by Eq. (3.3.2):[165]

$$\overline{I} = \overline{L} \times \overline{C} \qquad\qquad (3.3.2)$$

Average capacity is defined as the average rate that work is output by the system, and average manufacturing lead time is defined as the average total time required to produce a product from the start of the gating operation until the completion of the final operation. Lead time from receipt of a customer order until customer delivery is referred to as the supplier response time (SRT). Eq. (3.3.2) may be further developed as shown by Eq. (3.3.3):

$$\bar{I} = \bar{L} \times \bar{C} = \frac{\sum_{i=1}^{j} n_i t_i}{\sum_{i=1}^{j} n_i} \times \frac{\sum_{i=1}^{j} n_i}{T} = \frac{\sum_{i=1}^{j} n_i t_i}{T} \qquad (3.3.3)$$

Where:

j : Number of lots per period of analysis
n_i : Number of products in the ith lot
t_i : Manufacturing lead time of the ith lot in days
T : Number of work days in the analysis period.

The calculation of average inventory using Eq. (3.3.3) is best demonstrated by example. Consider Table 3.3.1. The average inventory calculation is shown by Eq. (3.3.4):

Table 3.3.1

Lot identifier	Lot size (n_i)	Work days per lot (t_i)	$n_i t_i$
X	20	4	80
Y	30	6	180
Z	50	10	500

$$\bar{I} = \bar{L} \times \bar{C} = \frac{760}{100} \times \frac{100}{20} = 38 \; units \qquad (3.3.4)$$

If the lot sizes are halved, the work days per lot will also be reduced by half. The X, Y, and Z lots would have to be iterated twice to equal the 20-day lead time horizon of Table 3.3.1. Reducing lot sizes by half results in an average inventory of 19 units. The relationship between a 50 percent lot size reduction and average inventory based on the data in Table 3.3.1 is shown in Fig. 3.3.5.

Clearly, the reduction of lot size results in a consequent reduction in lead time that significantly reduces inventory while maintaining the same capacity level. Lot sizes are a function of the value-adding processing time. A reduction in the non-value-adding component of production lead time will also reduce the average inventory level. The non-value-adding components of manufacturing lead time include setup time, inspection time, machine breakdown time, defect repair time, interprocess delivery time, and queue time.

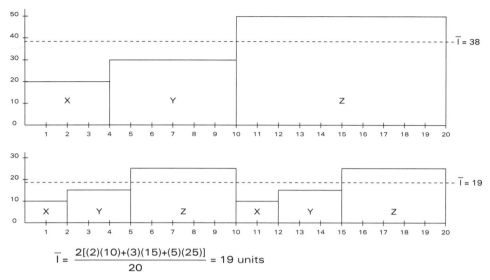

$$\overline{I} = \frac{2[(2)(10)+(3)(15)+(5)(25)]}{20} = 19 \text{ units}$$

Figure 3.3.5 Lot size and inventory.

Consider the case where the average inventory for a particular manufacturer is 240 units, the manufacturing lead time is 12 days and the average capacity is 20 units per day as shown by Eq. (3.3.5):

$$\overline{C} = \frac{\overline{I}}{\overline{L}} = \frac{240 \text{ units}}{12 \text{ days}} = 20 \text{ units/day} \qquad (3.3.5)$$

A reduction of four days in the non-value-adding component of manufacturing lead time will result in an average capacity increase of 10 units per day. Thus, a 33 percent reduction in non-value-adding lead time increases capacity by a remarkable 50 percent. Alternative approaches to increasing average capacity include purchasing additional capital equipment, increasing the work force, and outsourcing. The problem with purchasing additional capital equipment and increasing the work force is that both approaches address the value-adding component rather than the non-value-adding component of manufacturing lead time. Outsourcing will reduce the ability of a manufacturer to recover overhead costs.

The reduction of non-value-adding time is predicated by a relentless search for waste. There are many symptoms associated with poor time management such as poor quality, rampant expediting, and inventory write-offs. Scheduling alone will not yield the lead time reductions necessary to obtain competitive advantage. As a result of inventory excesses occurring based on the non-value-adding time material spends in the system as WIP, it is essential to examine how organizational energy is being used throughout the manufacturing system. MRP compensates for worse than optimal value-adding time by inflating order quantities and this is one of the major shortcomings of MRP. MRP must not become an opiate for ignoring the reduction of non-value-adding time.

A physical on-hand inventory analysis should take place to measure the percentage of inventory that is non-value-adding. Non-value-adding is categorized for on-hand inventory as follows:

- *Idle:* excess on-hand inventory as a result of insufficient demand
- *Obsolete:* on-hand inventory that has been replaced or phased out due to a design change, specification change, or new product introduction
- *Damaged:* on-hand inventory not fit for use
- *On-hold:* on-hand inventory that does not meet specification
- *Rework:* on-hand inventory that can be salvaged through repair.

The three logical groupings for on-hand physical inventory are raw material, WIP, and FGI. A focused analysis based on product, inventory value, or other logical subdivision should be performed to identify non-value-adding inventory and facilitate root cause analysis. Inventory that is non-value-adding contributes to the cost of doing business and must be eliminated.

The inventory turnover ratio (i.e., manufacturing cost of goods sold / average inventory investment) assumes that all inventory is equally likely to "turn" in the short term. This assumption is invalid. While most non-value-adding inventory will never turn over, a portion of the total non-value-adding inventory base will turn over on a quasi-regular basis (e.g., rework). This is an indication that overbuying and overbuilding are most likely occurring. In order to compensate for less than optimal quality levels, the master plan is often inflated to ensure delivery dates will be achieved. This will tend to hide rather than expose problems associated with non-value-adding inventory. Inventory that is non-value-adding consumes valuable working capital and the cost of carrying such inventory is unknown by most companies. The most significant problem associated with zero turnover non-value-adding inventory is the risk and cost consequences of this inventory entering the manufacturing system.

The ratio of non-value-adding inventory to value-adding inventory (expressed in dollars) is a sound metric for measuring performance. Non-value-adding inventory distorts the true cost of the aggregate inventory investment and inflates labor, material handling, warehousing, insurance, and data processing costs. The tax burden for a company is also increased. Non-value-adding inventory must be eliminated and root cause analysis should be performed to prevent recurrence in the future.

In practice, it is rare for an inventory manager to deal with inventory planning problems uncontaminated by the uncertainty associated with a forecast. There is uncertainty in both the demand for the product and the delivery

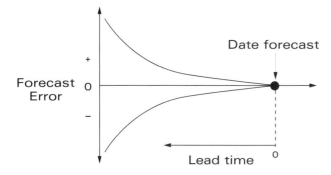

Figure 3.3.6 The relationship between lead time and forecast error.

dates of suppliers; however, the occurrence of a stockout is always a distinct possibility. By reducing manufacturing lead time, the influence of forecast error is significantly reduced. The relationship between lead time and forecast error is shown in Fig. 3.3.6.

A build-to-stock manufacturer can almost seamlessly become build-to-order if manufacturing lead time is sufficiently reduced over time. It is important to note that customers who place large orders may fail to exploit the consequence of reduced manufacturing lead time. The supplier response time may be longer than it should be. From an inventory investment standpoint, a customer who places an order for 200 units at the end of the month would be better off by accepting partial delivery of 10 units per day for 20 working days rather than a batch delivery of 200 units at the end of the following month. The likelihood that 200 units delivered to a customer in one batch are utilized the same day or several days thereafter is minimal for a customer who is a high-mix, low-volume manufacturer. Such delivery patterns are usually indicative of poor customer or supplier manufacturing methods or both. In order to exploit reduced lead times, cross-functional linkages must be established between manufacturing, suppliers, and customers.

The proper management of inventory plays a crucial role in manufacturing success. From a strategic management perspective, capacity and manufacturing lead time are the two major factors that affect a company's inventory position. With regard to financial planning, this relationship is an invaluable tool for determining the amount of capital investment required to support the average work-in-process inventory. The primary objective of inventory management is to achieve the business objectives of the company at the lowest possible inventory investment level. Marketing, finance, and manufacturing must be cross-functionally integrated in order to effectively achieve the inventory investment objectives of the company.

3.4 Responsiveness

In order to increase the total manufacturing output generated by a manufacturing system, the setup constraints must be evaluated. Setup constraints require that each process or machine is properly configured prior to performing a particular operation. Setup constraints are also referred to as *hard constraints* and are nonrelaxable in the context of ensuring the production schedule is validated. Other constraints, such as available capacity and due dates, are referred to as *soft constraints* and are relaxable when they cannot be satisfied. Available capacity can be enhanced through overtime, increasing the work force, and outsourcing. Due dates can be changed when performance to schedule is not as planned. The reduction of setup constraint time is a competitive weapon that results in a manufacturing system that is capable of providing the customer with a greater amount of product choice within a given time frame.

Consider Fig. 3.4.1. A fabrication operation produces four products (J, Q, K, and A) and delivers them to customer operations one through four respectively. The total demand placed on the fabrication operation is four each J, Q, K, and A, and the processing time for each product at each process step is one time unit. Three possible schedules are shown in Fig. 3.4.2 for a setup time equal to one time unit, and three possible schedules are shown in Fig. 3.4.2 for a setup time equal to two time units. The production lot size rule for all schedules is varied while the transfer lot sizes are always one. The schedules of Fig. 3.4.2a will arbitrarily represent the company while the schedules of Fig. 3.4.2b will represent the competitor.

What is immediately apparent about the schedules of Fig. 3.4.2 is the adverse effect that setup time has on available capacity. Consider schedule I for the company and schedule II for the competitor. Although both schedules have the same makespan *(33)*, the comparative responsiveness and inventory positions for these schedules are different. Schedule I will respond to all customers every eight time units, and schedule II will respond to all customers

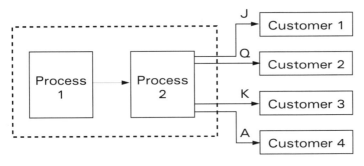

Figure 3.4.1 Fabrication operation.

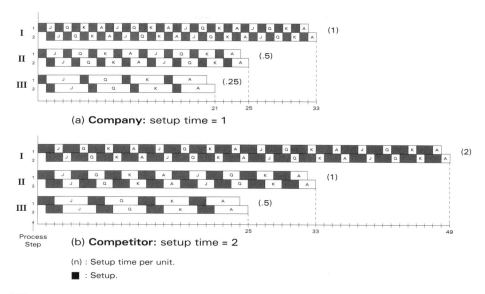

Figure 3.4.2 Gross versus net thoughput.

every 16 time units. The inventory position for schedule I is half the inventory position of schedule II. The customer response quantity produced (one unit per customer for schedule I and two units per customer for schedule II) is defined as *net throughput*, and the total quantity of products ordered and produced in a given makespan (four units per customer) is defined as *gross throughput*. The ratio of gross to net throughput is useful as a measure of responsiveness. Gross and net throughput may also be expressed in dollars.

In order to improve the competitor's inventory position and responsiveness, a reduction in lot size must occur. This can be accomplished by using competitor schedule I. Although responsiveness is improved from 17 to 12 time units, the makespan is significantly increased from 33 to 49 time units. If the competitor attempts to minimize makespan through the use of schedule III, it will be able to deliver all products faster than the company *(25 time units versus 33 time units)*. This performance comes at a very high price. The competitor's inventory position is significantly increased and responsiveness is decreased relative to the company *(22 time units versus 8 time units)*. In fact, the competitor is at a competitive disadvantage with respect to inventory position and responsiveness relative to the company, irrespective of any schedule chosen. The only way that the competitor can improve its ability to compete with the company is to reduce setup time.

Note that the setup times per unit for schedule I of the company and schedule II of the competitor are equal. Although the amortized setup time per unit for competitor schedule III is less than company schedule I (0.5 versus 1), the competitor's responsiveness and inventory position are inferior to the company's. Thus, the amortization of setup time is *irrelevant* when attempting to achieve a competitive advantage in customer responsiveness and inventory position. Management that rationalizes away large lot size production methods using adverse setup times as an excuse, is mortgaging the future competitiveness of their company if a setup time reduction project is not staffed and funded. Oftentimes a particular ordering of products to be produced will facilitate a reduction in setup time by minimizing the number of machine changeovers required. This strategy may result in the avoidance of a setup altogether for particular consecutively run products. A strategy of setup avoidance will dictate the order in which products must be processed and will severely damage the competitiveness of a high-mix, low-volume manufacturer. For high-mix, low-volume manufacturing environments, the processing times for products at each stage of manufacture are unequal, and the sequence in which products are processed should be based on process constraint time considerations. Schedules must not be developed based on local machine setup time avoidance criteria. Significant reductions in setup time can be achieved by simply performing as much setup preparation as possible while the machine is running (a.k.a. external setup) as opposed to performing these tasks while the machine is idle (a.k.a. internal setup). Rather than relying on engineering staff exclusively to reduce setup times, machine operators should be included in a setup time reduction project. By involving employees and using their knowledge, results and benefits will be achieved sooner. The benefits that result from reductions in setup times are increased capacity, improved customer responsiveness, and reduced inventory levels. The reduction of internal setup time will also minimize a source of manufacturing variability.

For prototype assembly operations, it is possible for the competitor to improve the responsiveness of schedule II in Fig. 3.4.2b without sacrificing the total time to deliver all products. The overriding concern of research and development (R&D) is to obtain performance feedback as soon as possible on a prototype product. If R&D places an order for four each J,Q,K,A, the faster that R&D can receive at least one unit of each prototype, the faster a potential performance problem will be detected. The remaining units of each product type are usually made available to manufacturing test engineering or others in the product development and evaluation process.

Consider schedule II of Fig. 3.4.3. Schedule II is a hybrid schedule that responds to the customer as quickly as possible with the first prototype unit for each product type, then batch-delivers the remaining prototypes. The difference in the responsiveness of schedule II in Fig. 3.4.2b as compared to schedule I in Fig. 3.4.2a is much greater than the difference resulting from the use of schedule II in Fig. 3.4.3 despite having a setup time twice that of schedule I of Fig. 3.4.2a. The gross throughput (i.e., makespan) performance of

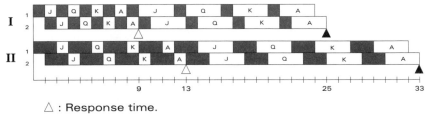

△ : Response time.

▲ : Makespan.

Figure 3.4.3 Hybrid schedules.

schedule I in Fig. 3.4.2a can be improved through the use of schedule I in Fig. 3.4.3 without sacrificing the required level of responsiveness. It is essential to understand that a company will not achieve the full measure of competitive advantage over a competitor with longer setup times if proper scheduling methods are not employed.

The relationships among setup time cost, inventory cost, and the total cost of responsiveness are depicted in Fig. 3.4.4a. The reduction of internal sequence-independent setup costs (i.e., internal setup time) is the strategic driver for improving manufacturing responsiveness performance. The effect of reduced internal setup times on the cost of responsiveness is shown in Fig. 3.4.4b. The cost of responsiveness improvement is maximized if lot sizes are inappropriately increased due to poor or nonexistent scheduling. There is a strategic advantage in knowing what the competition's setup times are. For competitors with comparable setup time performance, the company that does the better job of scheduling can achieve a competitive advantage in inventory position and customer responsiveness. The relationships among manufacturing, the competition, and outsourcing suppliers are critical when evaluating manufacturing responsiveness. Marketing, sales, R&D, and manufacturing must be cross-functionally integrated in order to benchmark and set standards for manufacturing responsiveness performance.

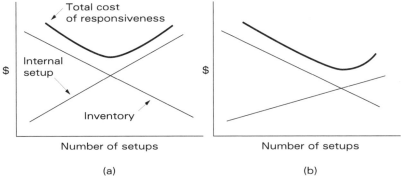

Figure 3.4.4 Responsiveness.

3.5 Balancing Capacity

One of the major factors affecting the efficiency of a high-mix, low-volume manufacturing environment is the inability to balance the amount of workload among the various resources required to produce the mix of products ordered. Balance delay equations can be used to measure the degree of imbalance that exists for the apportioned work content within a process step or across the entire production line. The benefits associated with a balanced production line are as follows:

- Increased efficiency
- Reduced variability
- Elimination of overloads and underloads.

Balanced flow lines are typically associated with repetitive Just-In-Time manufacturing environments that are competing on price. Although balance is *impossible* to achieve for a high-mix, low-volume manufacturing environment, the degree of imbalance can be minimized. The measure of production line balance is referred to as balance delay *(D)* and is mathematically expressed by Eq. (3.5.1) as the ratio of idle time to total available processing time:

$$D = \frac{\sum_{m=1}^{n} I_m}{\sum_{m=1}^{n} \tau_m} \qquad (3.5.1)$$

Where:

m : Machine or workstation number
n : Total number of machines or workstations required
I_m : Idle time at the mth machine or workstation
τ_m : Cycle time of the mth workstation.

The total idle time at a particular machine or workstation is expressed as:

$$I_m = \tau_m - \sum_{j=1}^{k} t_{mj} \qquad (3.5.2)$$

Where:

j : Job number
k : Total number of jobs
t_{mj} : Processing time for the jth job at the mth workstation.

By substitution we have:

$$D = \frac{\sum\limits_{m=1}^{n} \left(\tau_m - \sum\limits_{j=1}^{k} t_{mj} \right)}{\sum\limits_{m=1}^{n} \tau_m} = \frac{\sum\limits_{m=1}^{n} \tau_m - \sum\limits_{m=1}^{n} \sum\limits_{j=1}^{k} t_{mj}}{\sum\limits_{m=1}^{n} \tau_m} \qquad (3.5.3)$$

If we assume that all machines and workcenters are identical and note that the expression:

$$\sum_{m=1}^{n} \sum_{j=1}^{k} t_{mj}$$

represents the total work content time *(TWC)* for all machines or workstations, we can simplify Eq. (3.5.3) and obtain the expression:

$$D = \frac{n\tau - TWC}{n\tau} \qquad (3.5.4)$$

In order to obtain optimal balance, the idle time (i.e., numerator) must equal zero. The theoretical optimal number of machines or workstations required to obtain optimal balance is arrived at by setting the numerator equal to zero and solving for *n:*

$$n = \frac{TWC}{\tau} \qquad (3.5.5)$$

If the total work content time is not evenly divisible by the total available processing time for a particular machine or workstation, then a fractional machine or workstation would be required to obtain the optimal theoretical balance. In order to alleviate this problem, the total number of machines or workstations required is the smallest integer greater than or equal to the ratio of total work content time to total processing time. This is mathematically referred to as the "ceiling":

$$n = CEILING \left(\frac{TWC}{\tau} \right) \qquad (3.5.6)$$

Table 3.5.1

	t_1	t_2	t_3
A	8	10	15*
B	10*	5	8
C	15*	4	6
TWC	33	19	29
$n = CEILING\left(\dfrac{TWC}{\tau}\right)$	$\dfrac{33}{15} = 2.2 = 3$	$\dfrac{19}{10} = 1.9 = 2$	$\dfrac{29}{15} = 1.93 = 2$
$D = \dfrac{n(\tau) - TWC}{n(\tau)}$	$\dfrac{3(15) - 33}{3(15)} = .27$	$\dfrac{2(10) - 19}{2(10)} = .05$	$\dfrac{2(15) - 29}{2(15)} = .03$

* Denotes a capacity constraint

Consider Table 3.5.1. One each of products A, B, and C are produced by a serial flow production line that consists of three separate process steps. The processing times for products A, B, and C are t_1, t_2, and t_3 respectively and the optimal scheduling sequence is A, B, C (the MCS sequence) or A, C, B.

The longest processing time is associated with process step one *(33)* which is the bottleneck process step. The minimum cycle time for process step one is 15. The minimum cycle time for each process step is the maximum of that job step's processing times. Two decisions must be made in order to minimize balance delay. First, the optimal sequence must be determined and secondly, the appropriate cycle time must be established. Process steps one, two, and three will have balance delays of 27 percent, 5 percent, and 3 percent respectively, based on using the minimum cycle time. In the case where one machine or workstation is used at each process step, the balance delay for each process step would be zero, and the mix of products will be produced in the maximum practical cycle time. The number of machines or workstations required at process steps one, two, and three are three, two, and two respectively. For process step one, each product will be processed by a dedicated machine or workstation. For process steps two and three, products B and C will be processed by the same machine or workstation at their associated process steps while product A will be processed by a different machine or workstation. It is important to note that the nonconstraint process step two will be able to produce at a faster rate than the bottleneck process step and thus experience periods of starvation. If the rate of production for process step one is increased, a queue will form at the input to process constraint three. As a general rule, all nonconstraint process steps that feed constraint process steps should be paced by the bottleneck process step through the use of generic

Kanban signals as described in the previous chapter. For the case of Table 3.5.1, every effort should be made to increase the capacity of constraint process steps one and three in order to increase the overall throughput of the production line. Increases in capacity can be achieved by reducing or eliminating non-value-adding process time, purchasing more efficient equipment, redesigning the product, or improved scheduling methods. It is critical to understand that increasing the capacity of process constraint one or three alone will not significantly improve the overall performance of the production line.

If we consider the case of inexhaustible arrivals of products A, B, and C, the balance delay for process step one is considerably reduced over time by applying a parallel machine heuristic that will reduce the imbalance of work content times across the three machines or workstations. This can be accomplished by successively scheduling the longest remaining job to the machine where it will be completed the earliest. The balance delay at process step one will be reduced to 4 percent after the second iteration of the heuristic sequence is completed.

As a general rule, it is important to load the constraint process steps to maximum capacity and thus minimize balance delay. Only the constraint process steps should be loaded to maximum capacity. The ability to load the constraint process steps to maximum capacity is based on the lot sizes selected and the granularity of the processing times per unit of the products to be produced. If a nonconstraint process step is loaded to maximum capacity, a constraint process step will be overloaded. The constraint process steps limit the overall capacity of the production line. Any additional machines at process step two and/or three would be a waste and only result in decreased asset utilization levels, increased lead times, excessive inventory levels, and missed due dates.

Another application involving balance delay is the case where the optimal number of process steps for each product is selected with the result that the amount of work apportioned to each process step is as equally distributed as possible.

Based on Table 3.5.2, the number of process steps required to produce product C should be reduced from three to two. The result of making this change is shown in Table 3.5.3.

The optimal scheduling sequence is now A, C, B. The total number of units produced at process step two *(A+B+C=3)* per total work content time *(29)* is equal to that of process step three. Although an imbalance of work content time exits within the process steps, the imbalance of work content time across the process steps is significantly reduced.

The overriding goal of balance delay is to minimize the number of operators and WIP required to process a given amount of work at a given line speed and thus, optimize the utilization of available capacity. The choice of line

Table 3.5.2

	t_1	t_2	t_3
A	8	10^*	15^*
B	10^*	5	8
C	6	8	6
TWC	24	23	29
$n = CEILING\left(\dfrac{TWC}{\tau}\right)$	$\dfrac{24}{10} = 2.4 = 3$	$\dfrac{23}{10} = 2.3 = 3$	$\dfrac{29}{15} = 1.93 = 2$
$D = \dfrac{n(\tau) - TWC}{n(\tau)}$	$\dfrac{3(10)-24}{3(10)} = .2$	$\dfrac{3(10)-23}{3(10)} = .23$	$\dfrac{2(15) - 29}{2(15)} = .03$

* Denotes a capacity constraint

speed (i.e., cycle time) has a significant effect on the overall capacity requirements and should be carefully considered. Real-world line balancing optimization problems are generally solved using branch and bound methods. Branch and bound is a strategy by which a subset of the universe of feasible solutions is eliminated without having to explicitly evaluate them all. This approach significantly reduces computational requirements. Although a branch-and-bound strategy will not guarantee that an optimal solution will be found, the resulting balances are good with low balance delays. There are software applications available that work well for real-world line balancing problems. One particular line balancing system is called CALB (Computer

Table 3.5.3

	t_1	t_2	t_3
A	8	10	15^*
B	10^*	5	8
C	0	14^*	6
TWC	18	29	29
$n = CEILING\left(\dfrac{TWC}{\tau}\right)$	$\dfrac{18}{10} = 1.8 = 2$	$\dfrac{29}{14} = 2.07 = 3$	$\dfrac{29}{15} = 1.93 = 2$
$D = \dfrac{n(\tau) - TWC}{n(\tau)}$	$\dfrac{2(10) - 18}{2(10)} = .1$	$\dfrac{3(14) - 29}{3(14)} = .31$	$\dfrac{2(15) - 29}{2(15)} = .03$

* Denotes a capacity constraint

Aided Line Balancing) which was developed by the Advanced Manufacturing Methods Group at the Illinois Institute of Technology Research Institute.

Line balancing can offer a competitive advantage over competitors who indiscriminately schedule their production operations. The selection of cycle time (i.e., production rate) is critical for effectively balancing a production line and obtaining improvements in efficiency and capacity utilization. Marketing, sales, and manufacturing must be cross-functionally linked in order to effectively set the manufacturing production rate that will most efficiently, effectively, and profitably respond to the requirements of the marketplace.

3.6 Flexibility

For high-mix, low-volume manufacturing environments, it is essential to understand the relevant dimensions of flexibility, and the extent to which flexibility enables a manufacturer to competitively produce a variety of high-quality, low-cost products. Flexibility is defined as the capability or responsiveness to change. There are three intrinsic types of flexibility that are of general importance to a high-mix, low-volume manufacturer:

1. Mix flexibility
2. Volume flexibility
3. Work force flexibility.

Mix flexibility is the result of being able to build different types of products using the same production resources. The ability of the production line to absorb dynamic changes in product mix is a function of process design and is often referred to as process flexibility. Flexible processes are characterized by automation that has low sequence-independent setup cost per unit and manual processes that are prekitted (i.e., inventory is prestaged into assembly kits). Process flexibility is confounded with mix flexibility and should be thought of as a component of mix flexibility. The degree of mix flexibility required is based on a company's product strategy and the responsiveness of relevant competitors. Mix flexibility decisions must be made in the context of the cost involved to achieve a desired number of sequence-independent setups and production volume for a given investment in automation.

Consider Fig. 3.6.1a. The standard approach to evaluating an alternative investment decision is to use cost-volume, break-even analysis (CVBA). The total cost for each technology choice is equated and the resulting break-even volume is derived from Eq. (3.6.1):

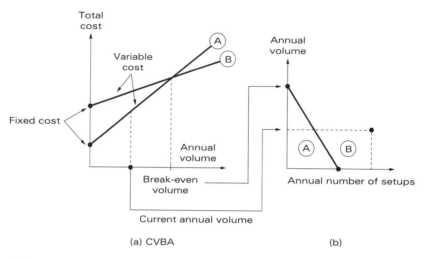

Figure 3.6.1 Process technology flexibility analysis.

$$(Fixed\ cost)_A + \left(\frac{Cost}{Unit}\right)_A \times Volume = (Fixed\ cost)_B + \left(\frac{Cost}{Unit}\right)_B \times Volume$$

$$(3.6.1)$$

If the current annual volume is not forecast to achieve the break-even volume in the future, the investment in new automation will not be made. The CVBA calculation does not take into consideration the setup cost associated with each automation choice. This shortcoming can have a devastating effect on the ability of a high-mix, low-volume manufacturer to enhance its competitiveness through increased flexibility. While the CVBA calculation derives the required break-even volume irrespective of the number of setups required, an analysis should be performed that determines the number of setups required irrespective of volume—derived from Eq. (3.6.2) (i.e., solve for volume then set volume equal to zero and solve for number of setups):[100]

$$(Fixed\ cost)_A + \left(\frac{Cost}{Unit}\right)_A \times Volume + \left(\frac{Setup\ cost}{Unit}\right)_A \times \#Setups =$$

$$(Fixed\ cost)_B + \left(\frac{Cost}{Unit}\right)_B \times Volume + \left(\frac{Setup\ cost}{Unit}\right)_B \times \#Setups$$

$$(3.6.2)$$

Consider Fig. 3.6.1b. It should now be obvious that the combined effects of current volume and number of setup requirements may result in the financial justification for the purchase of new automation. Although there are a variety of operations research and management science techniques that consider volume and number of setups criteria, the complexity of these approaches makes them less easily understood and communicated. The number of different products manufactured should not be used as a measure of mix flexibility. There are dissimilarities among the various products produced, and the ability to effectively make appropriate relative comparisons to the competition will be extremely difficult.

Volume flexibility is of importance in the context of process configuration. The benefits of volume flexibility are measured in terms of capacity utilization. The total number of possible sequences for n products and m parallel machines is $(n!)^m$. A decrease in the mix of products $(n-x)$ that can be processed on a particular machine $(m-y)$ will decrease mix flexibility disproportionately $[(n-x)!^{m-y}]$. Changes in production volumes based on demand changes will not be as effectively responded to if the number of sequences available to be processed on a particular machine is reduced. For example, the excess capacity that results from reduced demand on a particular machine cannot be used if an overload occurs on another machine as a result of excess demand. Group technology and volume-focused process configurations will reduce mix flexibility and result in a consequent reduction in volume flexibility. It is critical to understand that mix flexibility and volume flexibility are mutually reinforcing. The overriding focus must be on increasing the entropy (i.e., defocus) of the manufacturing system. Increased mix flexibility results in a consequent reduction in volume fluctuations that in turn increases volume flexibility. Volume flexibility can be enhanced through methods such as overtime, outsourcing, or hiring additional temporary workers. Employing these methods will result in adverse costs. Outsourcing decisions should not include the complete transfer of a product. The capability to produce products should be maintained in-house and at the virtual supplier (i.e., cosourcing). The virtual supplier will only produce when an excess capacity condition occurs. If a complete carte blanche transfer of a product is made to a virtual supplier, there is a risk of losing in-house expertise. *Cosourcing enhances flexibility while outsourcing does not.* Volume fluctuations can be measured absolutely by using simple variance measures. An effective method for measuring volume flexibility involves the total cost function for a manufacturer. The average of the total cost function will be relatively level under conditions of volume flexibility. Maximization of profits occurs when a manufacturer produces at the point where marginal revenue equals marginal cost, and this occurs at the lowest point of the average cost curve. In this way we can relate volume flexibility to the profitability of a manufacturer.

The work force represents the greatest source of flexibility. Virtually all complex tasks must be performed by the work force. The greatest contribution of the work force results from their problem-solving abilities. The work force should be well trained, cross-trained, and empowered to effect process improvements. The primary driving force behind a knowledgeable work force is the competence of their supervision. A technically trained and skilled supervisor is required to effectively develop a flexible, trained, motivated, and creative work force. Ultimately, the competitive advantage for any company is a direct result of the quality of its work force. A highly trained and flexible work force will enhance overall manufacturing capabilities. The work force should be focused on the customer and competition with an emphasis on standardization, simplification, and creativity as they relate to improving the *global* performance of the organization. The essential function of the work force is the never-ending pursuit of continuous improvement. Ideas generated from all levels of the work force cost nothing and are the only source of *positive* change. Managers must listen to and reward employees who contribute ideas that enhance flexibility and reduce complexity of existing resources.

A plethora of changes can occur independently of the manufacturing system such as vagaries of incoming order mix and volume as well as poor vendor quality. When these external changes are coupled with internal changes such as machine breakdowns, process defects, and work force processing time variabilities, the ability to supply exceptional value goods and services necessitates both proactive and reactive manufacturing capabilities. In the ideal situation, performance measures are invariant to change and the manufacturing system is defined as being robust. From basic economic theory, it is known that the profitability of a flexible manufacturing firm will be relatively uniform over a particular range of product mix and volumes. High-mix, low-volume manufacturers that associate poor financial performance with changes in product mix or volume are in reality inflexible. This is a classic example of confusing the symptom with the cause. Poor financial performance is often a key indicator of a high-mix, low-volume manufacturing environment that is inflexible in either proactively or reactively responding to change. In order to respond rapidly and effectively to internal as well as external changes, the proper use of an effective and efficient manufacturing information and control system such as MRP II is a necessity.

The adoption of flexible manufacturing systems requires cross-functional coordination between all functions of the organization. This will have a significant effect on how a company is managed. Increased marketplace linkages coupled with demands for increased delivery and responsiveness performance demand a management team that defines, measures, and places an emphasis on flexibility. In a flexible manufacturing environment, economies of scale and increasing product mix are mutually reinforcing. Most importantly, performance improvements resulting from increased mix, volume, and work force flexibility can create effective barriers against the competition.

3.7 Complexity

Complexity is most often associated with the design-related issues of a product that is composed of interconnected parts that involve some degree of intricacy to manufacture. Complexity is related to the number of discontinuities associated with a product that, in turn, directly affects the ability to produce the product. Complexity also has a direct effect on the operating reliability of a product. Complexity factor *(CF)* (Pugh[120]) has been defined and quantified as shown by Eq. (3.7.1):

$$CF = \frac{K}{f} \sqrt[3]{N_p N_t N_i} \qquad (3.7.1)$$

Where:

N_p : Total number of parts
N_t : Total number of unique types of parts
N_i : Total number of interconnections and interfaces
f : Total number of functions the product is expected to perform
K : Constant of convenience.

This approach has met with success for electronic equipment and is a measure that product designers can grasp at any stage of the design process. Complexity is most effectively avoided by focusing on simplicity. When differentiating good and bad design from a manufacturing perspective, it is essential to avoid problems and errors as early in the design process as possible. In fact, manufacturability issues should be initially addressed during the conceptual stage of the design through the component level, subsystem, and total system levels. During the conceptual design phase, it is adequate to simply consider major components when calculating N_p, N_t, and N_i, excluding small parts such as nuts, bolts, washers, etc. N_p, N_t, and N_i should be considered as separate criteria and always be visible and continuously monitored. The lower the complexity factor for a particular product under consideration, the lower its cost, the higher its quality, and the higher its reliability. Clearly, the application of a zero defects strategy for production problems mandates a superior design in terms of product manufacturability.

Enhancing manufacturability criteria (i.e., reducing the complexity factor) during the design process is often referred to as design for manufacturability and assembly (DFMA). One aspect of good DFMA involves component part standardization during the design phase. Sufficiently high quantities of standard component parts will encourage small lot size deliveries to point-of-use stock locations and thus minimize work-in-process inventory.

Manufacturability is achieved during the design of the product and not the production process. It is essential for manufacturing to formally document and communicate DFMA problems to research and development. In fact, a high-mix, low-volume manufacturer should train research and development in the guidelines and rules of proper DFMA for their particular factory environment. A videotape can be used to effectively show the manufacturing life cycle of a product from receiving to shipment with a special focus on critical processes and their relationship to DFMA. Good as well as poor examples of DFMA should be shown. Training should be performed for all research and development clients irrespective of whether they are captive, another division, or contracted.

The cost and degree of effort required to manage uncontrolled growth in the number of different part numbers can hamper a high-mix, low-volume manufacturer's competitiveness in the marketplace. Research and development engineers should acquire some level of direct manufacturing experience and be educated in DFMA techniques. In today's competitive global marketplace, a minimal competitive advantage or disadvantage can significantly affect a manufacturer's performance in the marketplace. The benefits of DFMA are numerous and the following list is not exhaustive:

- Reduced time to market
- Improved quality
- Reduced cost
- Increased reliability
- Reduced redesigns
- Reduced engineering change-orders
- Increased automated assembly.

Cross-functional linkages must be established between manufacturing, purchasing, and research and development in order to ensure that a low-cost, high-quality, and high-reliability product is designed to meet the needs of the marketplace and thus achieve a competitive advantage. Competitive advantages arising from superior product designs are difficult for competitors to counteract. Innovative product designs can result in patents that establish long-term competitive barriers.

3.8 Process Quality Improvement

Manufacturing process quality is essential for economically satisfying customer requirements. Whether quality problems occur at the machine-operator interface or at an exclusively manual process, the ability to solve quality problems rests with people. No machine has ever solved a quality problem. Providing the work force with the appropriate knowledge, training, tools, and techniques to effect continuous quality improvement is of paramount importance to achieving competitive advantage in cost, delivery, and responsiveness.

The causes of poor quality are generally categorized into two subgroups, common causes (85 percent) and special causes (15 percent). Common causes, which affect all workers, are attributable to the manufacturing system and only management can reduce or eliminate them. Poor worker training, poor documentation, and large lot size production methods are typical of the types of system problems that result in poor quality. Special causes are defined as those problems that the worker can reduce or eliminate. As common and/or special causes are reduced or eliminated, the percentages can fluctuate.

If workers are empowered to have control over the system in which they work, the process of continuous quality improvement takes on a new approach. It is essential that workers are given the power to find local faults in the system and to take the appropriate action on them. Decentralization of management's responsibility to find the local faults of the system is a genuine attempt on the part of management to correct the local faults of the system. Once workers are empowered to effect changes in the system within which they work, a quality system based on traceability can be established. A quality system based on traceability can improve the learning and problem-solving processes used by the workers.

The goal of any quality improvement program must be customer satisfaction as perceived by the customer. The figure of merit for a manual assembly process is based on attribute data, and zero defects is the goal. Nonstatistical tools can directly affect quality improvement. When used in conjunction with statistical methods, the process of continuous quality improvement is greatly enhanced. An important tenet of a successful quality improvement program is traceability. Effective problem-solving tools and development and training material can be put in place for individual workers when traceability is possible. For manual processes, traceability is the ability to identify the individual worker associated with a defect. In spite of the aversion that many managers and workers have toward traceability, it can be made to work for workers rather than against them. *The primary purpose of traceability is to focus on workers who exhibit a superior level of performance.* There is a wealth of knowledge possessed by this subgroup of workers that must be discovered and used. What constitutes a "superior" level can only be defined based on reliable data that establish a statistically significant difference. It is counterproductive to reward workers with relatively high performance or punish workers for low performance if the difference among workers can be explained simply as random variation. Abuse or intimidation of workers with defect data cannot be tolerated. Management must establish trust and drive out fear to gain the support of the work force. Without knowing when, where, and with whom a quality problem originated, the ability to solve quality problems is significantly reduced. If workers must be rated on their quality performance, all workers within the workcenter should receive the same rating. Such an approach will foster teamwork and synergy during the process of quality improvement rather than making it competitive.

Traceability demonstrates to the work force that the company is serious about quality improvement and has a genuine interest in them as individuals. As a result of workers knowing how they are doing, they will be able to answer the question: "What can I do to improve quality for my workcenter and thus help to achieve the goals of my company?" Workers have an innate desire to do high quality work. To think otherwise is ludicrous. Clearly, you cannot hold workers responsible for what they cannot control. Traceability empowers workers and is a catalyst for realizing effective and efficient problem solving. Traceability should become a part of the culture of any manufacturing operation serious about total quality control (TQC).

Although the following example illustrates how quality improvements are achieved based on traceability data for an electronic printed circuit board (PCB) manufacturer, the method applies to any manual assembly process. Defect data collected for a one-month period by internal process customer operations can be represented on the PCB component locator drawing (Fig. 3.8.1a). Technique errors as well as several inadvertent errors can be identified. Fig. 3.8.1b represents the physical location of the inadvertent errors and Fig. 3.8.1c represents the physical location of the technique errors. Inadvertent errors are those defects that occur due to workers "not paying attention" and are predominately unintentional. Inadvertent errors occur with no discernible pattern. The types of errors that occur are unique to each worker. There is no common denominator among workers. With inadvertent error, no one knows who will make the error, what the error will be, or when the error is going to occur. When monitored over time, the chaotic nature of inadvertent error usually results in Pareto charts being constructed without reference to them. Many times they are lumped together in a category called "other." Unless a process is mistake-proofed, inadvertent error is seldom, if ever, focused on or solved. Technique errors are those defects that occur when workers lack the necessary skills or techniques to perform the job. Technique errors are relatively constant over time and affect the majority of workers in the workcenter. The majority of workers may also have identical errors. A common

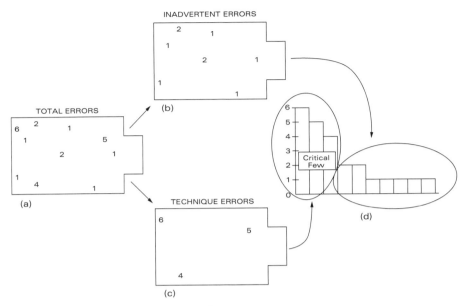

Figure 3.8.1 Classification of defects.

denominator exists among workers. Technique error can be caused by inadequate training of the workers, incorrect documentation, or management trade-offs. When productivity is made the critical managerial priority, quality is often the trade-off. Workers may employ adverse techniques to increase their productivity. Technique errors are therefore inadvertently management caused. Technique errors or inadvertent errors can represent the largest bars on a Pareto chart. The largest bars on a Pareto chart are commonly referred to as the critical few (Fig. 3.8.1d). Problem solving is almost exclusively directed at the critical few problems.

Tools are essential that help the workers focus on where they need to pay particularly close attention to what they are doing. Once traceability is established, all defect data should be supplied to the individual workers associated with the defects. Upon receiving the data, the workers can indicate the defects on the PCB component locator by drawing the physical location and number of failure occurrences for each defect detected. For high-mix, low-volume manufacturing environments, the assembly documentation should be displayed by computer screen to the worker. Documentation is then customized for each worker based on his or her associated defects. During subsequent component insertions, the workers will have a tool that can be quickly referenced to help them prevent future error occurrences. PCB component locator defect charts should be re-initiated at flexible time intervals to avoid cluttering the charts with data. The workers can develop their own Pareto charts (Fig. 3.8.2). Over time, workers will be able to monitor their personal

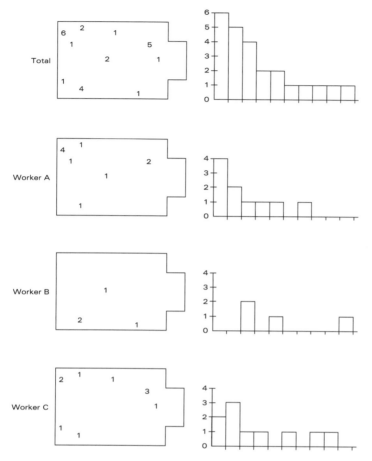

Figure 3.8.2 Total errors.

progress toward self imposed goals. Their goals will often be above the expectations of management. Individual worker improvements will result in work-center team improvement. It is critical to realize that an individual and team are not conflicting statements. Statistical charts, such as the u-chart or c-chart, should be used by the workers to follow their progress.

Inadvertent error is dynamic. There is a difference in fault spectrum between workers and fault spectrum shifts for individual workers. Fig. 3.8.3 represents the subgroup of total error that is inadvertent error. In most cases, quality circle environments cannot effectively address inadvertent error, because there is no common denominator among the workers. When mistake-proofing the process is not feasible, inadvertent error is most effectively addressed by the individual worker. The basic problem-solving tools enable the workers to learn what it is that they need to do differently to improve their quality (Fig. 3.8.2). Additional tools such as templates can be used for

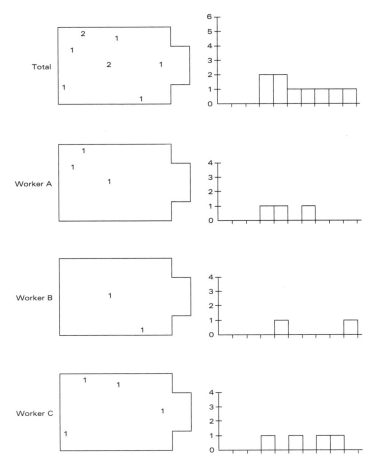

Figure 3.8.3 Inadvertent errors.

inspection purposes; however, this is not a tool to help workers do the job correctly the first time. Mistake-proofing the process is another possibility but can involve significant capital expense, particularly for present-day complex manual operations.

To effectively resolve technique error the performance of each worker must be analyzed over time. The time period must be long enough to establish sustained levels of performance. Fig. 3.8.4 represents the subgroup of total error that is technique error. Histograms can be used to show differences in performance among workers. Statistical methods should be used to determine whether the differences are significant. A subgroup of workers may exhibit a statistically significantly higher level of performance than the majority of workers in the workcenter (Fig. 3.8.5a). These workers typically have a knack for the job. A "knack" is the difference in technique that is responsible for the higher level of quality.

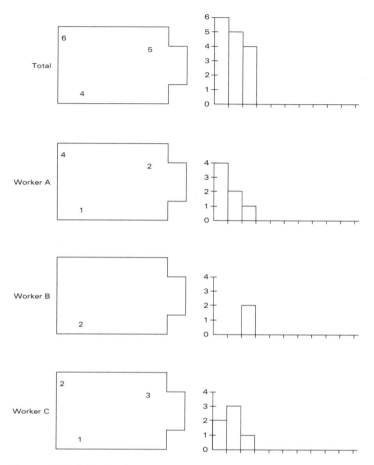

Figure 3.8.4 Technique errors.

Workers who have the knack may not know what it is that they are doing differently. An effective approach for uncovering beneficial techniques is to analyze the best and worst performers in the workcenter (Fig. 3.8.5b). Asking the workers how they perform the job or having the workers from these sub-groups use a flowchart to record how they perform the job to help uncover beneficial techniques. If this is not successful, the supervisor can watch the worker(s) on an asynchronous basis and discover the beneficial technique through observation. Quality improvements are likely to exist when there is a statistically significant difference in performance among workers in the work-center. However, techniques that yield improvements in efficiency may exist when there is not a statistically significant difference in performance among workers in the workcenter. A subgroup of workers may work smarter and not harder to achieve similar levels of quality, or the same worker might employ both good and bad techniques. In the latter case, performance as measured by

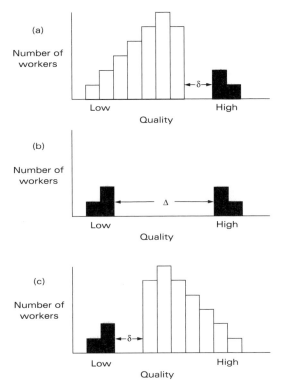

Figure 3.8.5 Problem-solving tool for technique error.

the number of defects might seem similar for workers not possessing the best technique(s). It can be very helpful to separate "technique" from "worker" by conducting experiments designed to test for differences among techniques independent of workers. An added consideration that can be easily handled in a designed experiment is identification of the possible interactions of competing techniques with downstream production processes. Clearly, the preferred technique(s) should not adversely affect subsequent processes. Once a knack is discovered, all workers in the workcenter should be trained in its use. This is an analytical approach for establishing standards of best practice.

A subgroup of workers may exhibit a statistically significantly lower level of performance than the majority of workers in the workcenter (Fig. 3.8.5c). Defects associated with this subgroup of workers are primarily management controllable. Workers may use adverse shortcuts when they feel management cares more about productivity than quality. This is particularly true when the reward system for the work force is based on productivity exclusively. Workers optimize their performance based on the reward system. Quality establishes a company's success or failure and must be the top priority of every worker. Management must clarify this for the entire work force and establish

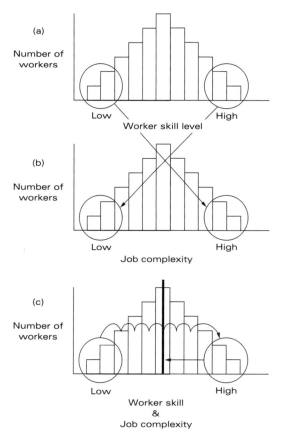

Figure 3.8.6 Worker flexibility.

the appropriate reward system to support it. In rare instances workers may willfully cause defects due to grievances against the company for unfavorable management actions. Improved management communication is the key to eliminating worker grievances. Management must understand grievances from the worker's vantage point and resolve the differences in conclusions based on the *facts* of the situation. If the performance of the worker does not improve after counseling by management, stronger options can be considered. Physical limitations such as color blindness can also result in poor quality.

It is the company's responsibility to achieve optimum results through the effective utilization of the work force. A quality system based on traceability is an invaluable tool to use for obtaining data to aid in the decision-making process. For a flexible work force where workers are required to perform several different processes, a mismatch between the worker and the process can have adverse consequences. Highly skilled workers assigned to low complexity processes can feel mistreated and become poorly motivated to do quality work. Less skilled workers assigned to high complexity processes can become

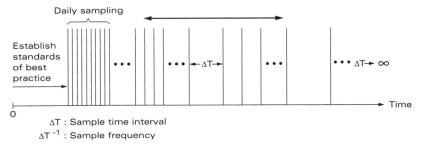

Figure 3.8.7 Sample frequency life cycle.

frustrated and produce low quality work when they are required to produce beyond their abilities (Fig. 3.8.6a and b). Less skilled workers should be developed in incremental steps to perform increasingly complex tasks. Highly skilled workers should not be assigned to processes of less than moderate complexity (Fig. 3.8.6c).

Fig. 3.8.7 represents the sample frequency life cycle for a quality system based on traceability. Until standards are established and workers are trained, feedback of data to individual workers should not take place. After standards are established and the process is changed to reflect "best practices," it will take a period of time to determine whether the process is in statistical control. The workers should begin receiving defect data on a daily basis and control-chart their individual performance. In most cases an automated defect data tracking and reporting system should be used. The defect data should be supplied to the individual workers by their immediate supervisor. Supervisors must provide leadership for the quality system to be efficiently and effectively implemented. Due to the learning curve, it is likely that the new process will initially be out of control. The workers should address the special causes as needed. Once the process is in control and only common causes exist, workers should resist the temptation to react to individual data points. This variation is an inherent part of the system within which they work. Preferably, management and labor should team up to identify and eliminate the sources of variation in the system. These sources can include product design problems, improper product specifications, inadequate training of workers, or poor working conditions.

Over time, quality improvements will change the frequency of fault occurrences. A reduction in the frequency of fault occurrences can be accompanied by a corresponding reduction in the defect sampling frequency. The defect sampling frequency should be increased as much as possible during periods of experimentation. An important part of any discrete sampling system is the sample frequency. For efficient learning and problem solving, daily feedback is essential. Defect data will be delayed, however, by the cycle time for downstream processes that discover and report defects. Small lot size production methods are critical in this regard. When costs associated with defects are suf-

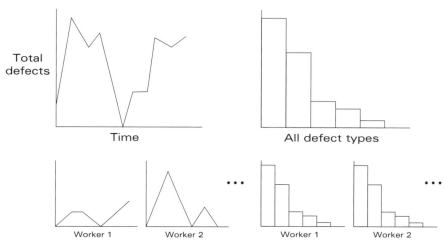

Figure 3.8.8 Problem-solving tools.

ficiently high, introducing an inspection step within the workcenter can accelerate the problem-solving process. The sample frequency will approach zero for continuous quality improvement. For the visionary goal of zero defects, the sample frequency is zero (i.e., no sampling at all).

Quality teams or workcenter teams whose charter is quality improvement typically monitor quality metrics on a weekly basis with two types of charts. One chart monitors the overall number of defects (e.g., u, c, or p charts) for the workcenter and the other is a Pareto chart that monitors the critical few fault spectrum for the workcenter. Brainstorming the possible causes of worker error using the Ishikawa[73] diagram (a.k.a. fishbone diagram) is relatively commonplace. Workers should develop or generate through computer automation their own personal statistical defect chart and Pareto chart to know what their contribution is to the overall results of the team (Fig. 3.8.8). By comparing their fault spectrum with that of the team, the workers will discover if they have a knack. A statistical test can determine if the difference is significant. If the knack is due to a beneficial shortcut, the worker should be asked to present it to the workcenter. Workers at lower performance levels can leverage the new technique when they perform the job in the future. It is important that the workers know when they have something valuable to contribute. Without this knowledge, workers who are using beneficial shortcuts will not contribute them.

Without a quality system based on traceability, individual workers simply cannot know how well they are doing. Problem solving without traceability is based on guesswork and worker theories. Rather than making quality improvement a certainty, uncertainty prevails. A quality system without traceability is an unfair hindrance to the worker. With traceability, brainstorming possible causes for defects is transformed into discovering the existing methods and techniques

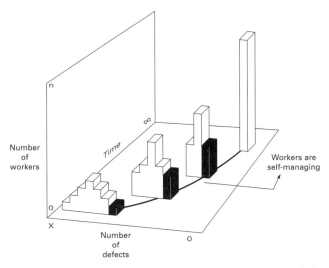

Figure 3.8.9 The continuous improvement model.

that facilitate the prevention of defects. Once traceability is implemented, a rapid decrease in the defect level can occur. The magnitude of the decrease is a function of the performance level of the best workers. The knowledge necessary for initial improvement already exists. The rate of improvement is determined by how quickly this knowledge can be discovered and subsequently adopted by all workers in the workcenter. Daily feedback of defect data facilitates effective and efficient learning. After the learning curve matures for workers adopting the new technique(s), the difference in performance among workers will also decrease. Further improvement will now become more difficult and time consuming. The knowledge required for further improvement does not exist among the workers alone. Designed experiments should now be used to effectively test new ideas generated by all levels of the manufacturing organization. Improvements beyond this level are extremely difficult to attain. Technology such as robots may eliminate the residual defect occurrences; however, the return on investment or processing time may be unacceptable. The technology will not pay for itself due to insufficient defect levels or increased processing time. The workers are now the prime source of added value.

When traceability has become a part of the company's culture and the workers have become proficient in the use of the problem-solving tools, the workers should be empowered to self-manage quality improvement for their workcenter (Fig. 3.8.9). Quadratic improvements in the cost of quality can be realized. Figure 3.8.10 represents the loss function for a quality system based on traceability.

Total quality control is not a spectator sport. Management must provide effective leadership. The focus of that leadership must be on understanding internal as

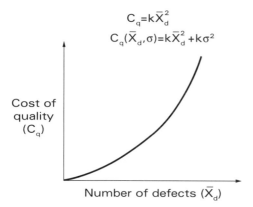

$$C_q = k\bar{X}_d^2$$

$$C_q(\bar{X}_d, \sigma) = k\bar{X}_d^2 + k\sigma^2$$

Cost of quality (C_q)

Number of defects ($\bar{X}_d$)

Figure 3.8.10 Quality loss function.

well as external customer needs, planning for the achievement of quality goals, and following up with periodic reviews. The organization that wants continuous quality improvement must be willing to pay the price, not in dollars, but in the operational commitment that this entails. Management must consider how a quality system based on traceability relates to the company's capabilities. Here, management is not only involved with factors that affect manufacturing, engineering, total capacity, intrinsic capability, and capitalization, but also with less obvious aspects such as overall economics, support effort, administrative requirements, effect on overall profit, strategic modeling, and underextension versus overextension. Experience has shown that a quality system based on traceability significantly improves the morale, confidence, and job satisfaction of the workers. As quality and worker satisfaction improve, the manufacturing process will run more smoothly, productivity will increase, on-time delivery performance will improve, inventory levels will be reduced, and costs will decrease.

Most companies tout their people as being their most important asset when, in reality, their management system causes them to be a liability. This can be fundamentally remedied by implementing a quality system based on traceability. Trial and error quality improvement efforts must be abandoned. Middle managers must encourage supervisors to believe that they can make a difference. Through active participation in the quality program, both quality and worker satisfaction will greatly improve. Middle management must convince supervisors that such efforts will be rewarded. It is important for managers to realize that there is not a utopian quality improvement system. Even the best quality system, once implemented, develops problems and difficulties, taking on a consumer life cycle of its own. One must keep track of what is learned about a quality system after it is introduced and implemented. Such knowledge can lead to strategic quality system modification. What emerges is a commitment to *comprehension* of total quality control.

3.9 Product Costing

A business is required by law to use generally accepted accounting practices (GAAP) for external reporting purposes. Under GAAP rules, products are valued at full cost based on material and labor directly traceable to a particular product and a percentage of indirect manufacturing overhead cost. A significant percentage of overhead cost is not traceable to particular products (for instance, general manager, legal personnel, support staff, janitors, etc.). Overhead costs are typically allocated based on direct labor hours or machine hours—volume related measures. The basic underlying assumption of a cost accounting system that allocates overhead from an aggregate cost pool in this manner is that overhead costs vary with the volume of products produced. The consequences of this "bushel basket" cost accounting strategy are devastating to a high-mix, low-volume manufacturer.

A peak of incoming demand will often occur at the end of the time period for which sales quota performance is measured. The result is a "hockey stick" demand pattern (Fig. 3.9.1a). Improper scheduling techniques are often employed to transform this demand pattern into a leveled demand pattern based on the *number* of units produced per period (Fig. 3.9.1b). The allocation of overhead based on volume creates the problem of period-end absorption. At the end of a period (e.g., monthly) either a sufficient quantity of products is produced to absorb the desired level of overhead or an insufficient quantity of products is produced with a negative overhead variance being subsequently reported by the finance department. There is intense pressure on managers to meet shipping targets based on revenue. In order to achieve shipping targets, the infamous end of the month push to ship products occurs, and more often than not, overtime and expediting are employed. The result is an increase in manufacturing cost and a monthly "hockey stick" shipment pattern (Fig. 3.9.1c). The problem is further exacerbated by basing performance improvement goals for the future on historical performance levels. Manufacturing is thus managed by quantity rather than by cost. This is a classic example of management not knowing what they are doing and having a cost accounting system to help them. Although aggregate measures are acceptable accounting practices for external reporting purposes (that is, balance sheet, income statement) as taught in college level business administration curricula, they are wholly inappropriate for managerial decision making at the operational level.

Labor productivity becomes the central focus for improvement efforts when using standard cost accounting approaches to allocating overhead. Labor productivity does not take into consideration overhead costs even though labor costs are minimal compared to the costs of materials and overhead. Under standard cost accounting rules, overhead is not directly linked to

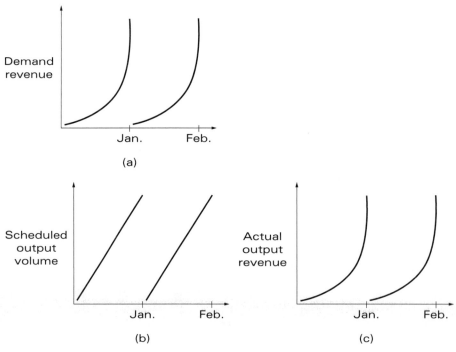

Figure 3.9.1 Hockey stick demand and shipment pattern.

the department that generates it. If a department reduces overhead, the resulting savings will be spread like peanut butter across the entire organization and show little benefit. Since departments are not linked to the overhead they use, the demand for overhead resources such as product engineering and process engineering is increased. Engineering resources are often used by departmental management to improve, you guessed it, labor productivity. The entire manufacturing organization is now driving down labor costs and at the same time increasing overhead costs. Rising costs in the face of improved labor productivity present a quandary to upper level managers. In order to address the problem of rising costs, upper level managers will, you guessed it, hire more managers. It gets worse. If labor productivity improvements stagnate, the work force payroll also stagnates and upper level managers may begin to look worldwide for lower wage rates and consider the transfer of their operations off-shore.

Outsourcing a portion of the manufacturing base is another consideration. The problem with outsourcing is that outsourced products will have zero allocation under standard cost accounting rules. There will now be fewer in-house products among which overhead can be allocated and the resulting cost increase will only serve to further fuel outsourcing decisions. Although outsourcing makes sense under the condition of insufficient in-house capacity, it almost never makes sense based on the rationale of achieving lower product cost. Standard cost accounting methods will incorrectly cost products. This is particularly true for a high-mix, low-volume manufacturer that produces low-volume, high-complexity products. Low-volume, complex products typically require intensive planning, setup, inspection, test, and engineering overhead support. Overhead allocation under standard cost accounting rules is based on volume considerations. The higher-volume, lower-complexity products will thus subsidize the low-volume, high-complexity products. Clearly, standard cost accounting methods are in direct conflict with the notion of "economy of scale" and will incorrectly cost high-complexity products. Predicted cost savings as a result of outsourcing higher volume products will therefore be flawed. Product pricing is based to a large extent on product cost. The over-pricing of high-volume products coupled with the underpricing of high-mix products can place a company at a competitive disadvantage if its competitors are properly pricing their products. This is particularly true if the high-volume products are priced approximately equal to or higher than the competitor's products. The associated loss of market share as a result of a flawed costing system is a very high price to pay.

Design for manufacturability and assembly (DFMA) efforts will not achieve their desired results under standard cost accounting methods. Standard cost accounting approaches will lead research and development engineers to believe that reduced product cost will result from specifying unique low cost components that minimize total material requirements. A proliferation in the number of unique components for any manufacturing environment results in increased overhead cost to plan, warehouse, document, and otherwise maintain the unique component part. The savings in parts cost may be overshadowed by the resulting increase in overhead cost.

Decisions that managers make must be tied to their effects on overhead costs. Activity based costing (ABC) serves this purpose. Activity based costing is based on the notion that products consume activities and activities consume resources. Multiple activity pools are established. Costs are collected for each activity (i.e., overhead and labor) and allocated based on their associated cost driver to the appropriate product(s) affected. Under the ABC method of cost accounting, the central focus is on activities, and these activities are managed to control and reduce costs. Business decisions based on ABC information are referred to as activity based management (ABM). An assessment must be made as to whether a particular activity (e.g., overhead, labor) is value-adding or non-value-adding. Activities should be improved or eliminated. Characterization of activities for the purpose of ABC should be led by a manager highly competent in how his or her particular manufacturing environment functions, and how it should be functioning, in order to effectively differentiate between value-adding and non-value-adding activities.

The method used to calculate the total cost associated with the in-process production operations for a particular product is referred to as *velocity costing.* The cost of in-process products varies according to the time-based accumulation of material, labor, and overhead costs at each process step. The average in-process cost is calculated using Eq. (3.9.1):[165]

$$\bar{v} = \alpha v_f \qquad (3.9.1)$$

Where:

$\bar{v}$: Average per-unit in-process cost
α : Cost accumulation factor
v_f : Final accumulated cost.

Consider the arbitrary cost accumulation pattern shown in Fig. 3.9.2. The general expression to derive the cost accumulation factor is given by Eq. (3.9.2):

$$\alpha = \frac{\sum\limits_{p=1}^{n} (v_p + v_{p-1}) t_p}{2 v_f \sum\limits_{p=1}^{n} t_p} \qquad (3.9.2)$$

Where:

n : Total number of process steps
t_p : Processing time at the pth process step
v_p : Accumulated cost at the end of the pth process step
v_f : Final accumulated cost.

The cost accumulation factor for Fig. 3.9.2 is calculated as shown by Table 3.9.1 and Eq. (3.9.3):

Table 3.9.1

p	t_p	v_p	$(v_p + v_{p-1}) t_p$	
0	0	20	20 (0) =	0
1	4	40	60 (4) =	240
2	3	100	140 (3) =	420
3	5	150	250 (5) =	1,250
4	3	200	350 (3) =	1,050
	$\Sigma = 15$		$\Sigma = 2,960$	

NOTE: v_0 = Material and overhead cost prior to the initiation of process step p_1.

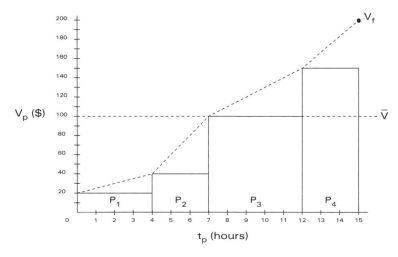

NOTE: P_0 = Material cost prior to initiation of process step P_1.

Figure 3.9.2 Manufacturing process cost accumulation factor.

$$\alpha = \frac{2960}{2(200)(15)} = .49 \ \ Therefore \ \ \bar{v} = \alpha v_f = .49(200) = \$98.67 \approx \$100 \ per \ unit$$

$$(3.9.3)$$

The derivation of the total working capital required is developed by first recognizing that average work-in-process inventory level is controlled by average capacity and average manufacturing lead time Eq. (3.3.3). By weighing average capacity and manufacturing lead time monetarily, the general mathematical expression for the working capital requirement (W_1) can be developed as shown by Eq. (3.9.4):

$$W_1 = \overline{V_L} \times \overline{V_C} = \frac{\sum_{i=1}^{j} n_i t_i \overline{v_i}}{\sum_{i=1}^{j} n_i \overline{v_i}} \times \frac{\sum_{i=1}^{j} n_i \overline{v_i}}{T} = \frac{\sum_{i=1}^{j} n_i t_i \overline{v_i}}{T} \qquad (3.9.4)$$

The working capital requirement for the example of Table 3.3.1 is calculated as shown by Table 3.9.2 and Eq. (3.9.5):

Table 3.9.2

Lot identifier	Lot-size (n_i)	\$/Unit $(\bar{v}_i)$	Work days per lot (t_i)	$n_i\, t_i\, \bar{v}_i$
X	20	30	4	2,400
Y	30	70	6	12,600
Z	50	50	10	25,000

Note: \$/Unit is given

$$W_1 = \overline{V_L} \times \overline{V_C} = \frac{40,000}{5,200} \times \frac{5,200}{20} = 7.69 \; days \times \$260/workday \approx \$2,000$$

$$(3.9.5)$$

The average working capital (\$2,000 per month) will turn over every 7.69 days. Value-adding manufacturing lead time is lengthened in real-world manufacturing environments due to delays caused by power outages, machine breakdowns, material shortages, interprocess transfer time, inspection time, and quality problems. The result is an increase in manufacturing cost, and an ongoing effort should be funded to reduce or eliminate these deficiencies. If the percentage (fractional) of time that interruptions take up is known and statistically characterized (ρ), the value-adding manufacturing lead time (L) should be modified as shown by Eq. (3.9.6):

$$L_{modified} = (1+\rho/100)L \qquad\qquad (3.9.6)$$

It is critical to understand that lead time measures should be representative of the time it takes for a manufacturing production operation to negotiate the entire mix of products produced. The period of time for which lead time performance is evaluated is based on the lot sizing rule employed. Reduced lot sizes will yield corresponding reductions in production cost.

It is important to express lead time in terms of actual working days, excluding holidays or other occurrences that result in a plant shutdown, in order to more accurately calculate production cost. If manufacturing lead time is expressed in hours, it can easily be converted to days by dividing by the product of the number of shifts worked times the hours worked per shift. An approximate conversion to actual calendar days worked is a simple matter of multiplying the lead time expressed in days by the total number of calendar days in one year *(365)*, and dividing the result by the number of actual working days for the year.

The primary barrier to adopting ABC in place of a standard cost accounting system is the difficulty of quantifying the cost of errors prevented by using ABC. The adoption of ABC requires that cross-functional linkages be established between marketing, research and development, finance, and manufacturing, in order to fully realize the financial benefits of ABC. All participants should be thoroughly educated in the fundamentals of ABC as well as the new paradigms in management decision *thinking* that will be required.

3.10 Work Measurement

Frederick W. Taylor[151, 152] is known as the father of scientific management and was responsible for the first definitive approach to work measurement. Soon after Taylor began his work, Frank and Lillian Gilbreth performed numerous detailed laboratory studies in motion and methods and developed what is known as micromotion study that forms the basis for modern day industrial engineering work measurement methods. Work measurement is a highly reliable way to achieve improved production methods and, most importantly, enables predictability in the timing of production operations. Predictability in timing of production operations is essential if customer delivery date commitments are to be met.

The application of expert industrial engineering knowledge in a systematic manner through scientific method is central to work measurement. The use of scientific method is based on the recognition that true knowledge about anything is based on facts. Work measurement is impractical if it is not integral with good management. Although the agents of production are people, money, materials, and machines, people are placed at the pinnacle since business exists by and for the social ends of people.

An industrial engineer assigned the charter of work measurement must combine the viewpoints of both engineers and workers in the field of human relations to a much higher degree than any other individual found in industry. Industrial engineers must also know how to maximize the performance of human resources in order to maximize production performance results. They must generally have skill in cooperating with and guiding the work force into better work habits. While this is the art of work study, the science is based on impersonal measurement techniques. The art and science of work study are equally important.

The variables that affect human production methods are functions of the effort, skill, and conditions under which an individual worker performs a given task or group of tasks. Work measurement is the most reliable approach to improving productivity and predictability of manufacturing operations. Standards based on work measurement methods should not be established until the work under consideration has been simplified as much as possible. Processing times based on historical performance will hide rather than expose poor work methods and their resultant inefficiencies. Nowhere are the benefits

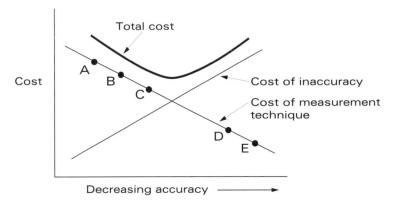

A. Stopwatch timing

B. Work sampling

C. Multiple regression analysis

D. Budgetary standards

E. Historical standards

Figure 3.10.1 The cost of work measurement.

of work measurement more pronounced than at bottleneck and constraint operations.

There are various techniques that are employed to perform work measurement such as stopwatch timing, historical standards, budgetary standards, work sampling, and multiple regression analysis. The cost versus accuracy relationship for these work measurement techniques is shown in Fig. 3.10.1.

The particular method chosen is a function of the required accuracy, nature of the work performed, and management objectives. Time standards are based on the concept of normal pace and industrial engineers have developed a reasonably uniform concept of normal pace. Industrial engineers have been trained to exercise proper judgment and statistics have been used to validate the accuracy of time studies for manual and machine-based production operations. The fact remains that work measurement is still a "rough" measure for manual production processes. The precision and tolerance of time standards are improved for automated production processes and setup time reduction requires an intensive focus on work measurement.

Confounded with work measurement performance is the learning curve. The learning curve is a representation of the experience of an individual worker for a given task. The basic learning curve model is shown in Fig. 3.10.2 and expressed mathematically by Eq. (3.10.1):

$$\tau = t_{min} + (t_\alpha - t_{min})e^{-\beta N} \qquad (3.10.1)$$

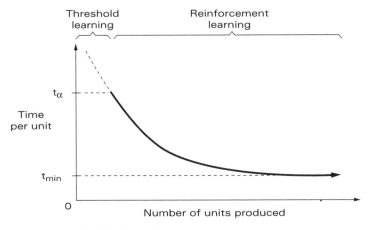

Figure 3.10.2 The learning curve.

Where:

τ : Time to perform a given task
t_{min} : Minimum processing time
t_α : Time until the first reinforcement cycle
N : Number of units produced
β : 1/N if t = t_α/e (i.e., only one unit is required).

Training of an individual worker is performed until the worker is able to perform a task unaided. This is referred to as *threshold learning*. Training in the conventional sense is over at this point and work measurement is not predictive during this period of time. Beyond the threshold learning phase, learning is referred to as *reinforcement learning* and is based on continuing experience and practice that strengthen the ability of the worker. Work measurement can be predicted at any point during this phase of learning.

Many behaviorists, industries, and universities have studied learning patterns extensively. Most researchers generally agree that a rapid reduction in processing time occurs during the initial phase of learning a new task, and the rate of improvement tapers off substantially as the number of units produced increases. In some cases, researchers have observed learning plateaus for certain workers.

The implications of the learning curve for a high-mix, low-volume manufacturer are clear. In spite of the low volumes, learning is primarily accomplished through proper task-specific training. Managers who blame poor quality or delivery performance on the inability of their work force to move up the learning curve, plainly and simply, do not know what they are talking about. In fact, such rationalizations will mask and discourage the search for the root cause of poor performance, and the status quo will continue to adversely affect production performance.

In the case where a temporary work force is used to augment a manufacturer's ability to increase capacity, there is an interesting learning phenomenon that takes place. When new workers are introduced into a department where workers have already been trained, the learning curve for the new workers is accelerated. New workers are driven to achieve similar or even superior levels of performance as compared to more experienced workers, based on competitive psychological factors. There is often a profound reduction in the time required to achieve significantly improved levels of performance when the new workers can leverage the knowledge and experience of the permanently employed workers.

When new products are introduced, the learning curve is most effectively achieved by training. However, if a new product introduction prototype production line is in place, it is critical to utilize the subset of the work force with the greatest intellect and talent. Workers will be required to learn new tasks on a continuous basis and not all workers may be up to this demanding challenge.

A manufacturing organization that institutes job specific training that is superior to its competitors' can achieve a competitive advantage in quality, cost, delivery, and responsiveness. Management, engineering, and the work force must be cross-functionally linked in order to effectively design training curricula that will align with the overall goals of the organization, meet the needs of the work force, and improve overall customer satisfaction measures.

3.11 Management Skills

Successful high-mix, low-volume manufacturing organizations do not just come into existence. Leading organizations are comprised of individually successful managers who lead people based on a vision of the future that is supported by clear objectives. These managers intensively focus on the needs of customers and build effective teams that are aligned with the overall goals and objectives of the organization. When considering the management skills required to successfully lead an organization, a significant gap between managerial skills and operational technical skills as related to high-mix, low-volume manufacturing environments can result in decisions that will make little or no sense on the factory floor. Eclectic approaches to addressing management problems often lack a coherent and theoretical basis and can damage an organization's competitiveness. Although scientific methods should not be the exclusive domain of managerial decision making, scientific methods have their greatest contribution in avoiding the implementation of fragmented and incompletely understood concepts. Plainly and simply, managers must have some level of detailed technical knowledge about what it is that they are trying to manage.[106]

The management of a high-mix, low-volume manufacturing environment can be studied and practiced deliberately and must not be a capricious, rule-of-thumb, seat-of-the-pants meandering from crisis to crisis. Organizations that spend millions of hours working on people behavior, activities, and attitudes are nothing more than incessant judgment machines. It does not necessarily follow that managers have sacrosanct people skills by virtue of their position. Although no organizational form is a panacea, the technical ability of a manager is essential for understanding the "core" of a business. Lack of technical knowledge can result in a management team that fails to fully make use of its assets.

Management that is exclusively focused on the province of people skills is often associated with the stereotypical notion that technically knowledgeable people have poor or nonexistent people skills. This stereotype is further fueled when managers who are deficient in technical skills have their decisions challenged by individuals within their organization who are technically proficient. Managers who lack technical knowledge are insecure in this knowledge and may lash out at scrutiny by attacking the style, motivation, or people skills of the messenger. Managers that feel exposed or threatened when their plans are scrutinized often attempt to criminalize this activity or simply write off the messenger as uninformed and lacking all the data. Upper management must work toward understanding the distortions created by such behaviors and set the standard by listening intently to alternative points of view. If upper level managers do not speak out against this sort of behavior, I might ask you, who will? Leadership demands putting aside hidden biases, preconceptions, hidden agendas, and stereotypes. Leaders must fully respect the ideas, contributions, and talents of their people based on their merits. Polarized functions within the management hierarchy that discourage cross-functional participative interactions for any reason must be eschewed.

It is important for managers to understand that workers scrutinize and criticize management the way lawyers criticize judges. The essential requirement is that a manager be scrupulously fair based on fact. The work force has the faculty by which to know right from wrong, and will conscientiously object to managers who create negative perceptions of employee behaviors that are characterized as being misguided political rationale or defensive. Indeed, political forces are often more bitterly resented. The work force does not impute nobility on managers strictly by virtue of position within the management hierarchy. Adverse managerial behaviors must be uncovered and are most effectively exposed by an anonymous survey of the employee base. Management must focus on how it treats its work force and place an emphasis on openness rather than servitude. Openness is a matter of survival and such an environment will foster a highly dedicated and participative work force that will not stand on the sidelines and throw up their hands and give up.

Empowerment of the employee base to contribute ideas during the decision-making process is critical to achieving improved performance. This is particularly true when decisions are to be made that directly affect the work they are doing. Managers who preclude action on the knowledge of their subordinates in favor of their own positional power lack the depth and character to be effective leaders. Managers of this nature seldom develop their people and can be further characterized as being averse to open and honest communication. In the most severe case, such managers will literally abdicate their work groups without supplying adequate guidance. Without adequate management skills, overall work group performance improvement efforts will be hampered. The negative impact of such managerial actions will have a rippling effect throughout the organization.

In 1924, Elton Mayo[93] performed experiments at the Hawthorne, Illinois plant of Western Electric Company that formed the basis of the human relations movement. The Hawthorne study demonstrated that the involvement of workers in planning, organizing, and controlling their own work results in an environment of positive, reinforcing cooperation when coupled with a close identification with management. In fact, there was a significant increase in productivity. The work force no longer looked upon themselves as isolated from the company. The relationships that developed elicited feelings of affiliation, competence, and achievement. The unsatisfied needs of the work group were now being fulfilled. The misinterpretation of what work force empowerment is and how it is effectively instituted will most likely damage a company's competitiveness. Abdication leadership denies workers the satisfaction of esteem and self-actualization and is referred to as laissez-faire leadership. The Hawthorne study further demonstrated that a lack of management leadership led to tension, anxiety, and frustration in the work force. Managers who institute changes and then subsequently abdicate further responsibility to their subordinates will create an environment where workers feel that they are unimportant, confused, and unattached victims of their own work environment. The findings of the Hawthorne study are just as valid today as they were at the beginning of the twentieth century. The only cited example of successful laissez-faire leadership involves the commonly cited case of the manager of a research and development laboratory. Subordinates in this case are highly skilled, highly educated, and well-trained professionals. Such professionals are typically highly focused and motivated and the job of a manager is to simply provide them with the equipment and support necessary to accomplish their work (i.e., administration). This example is the exception to the rule. In the overwhelming majority of cases a manager must carry out the critical function of a leader. The critical function of a leader is to influence and direct subordinates toward the attainment of goals which are in alignment with the overall goals of the organization. In order to effectively accomplish this, managers must effectively maximize and balance the technical and human relations aspects of what and whom they are managing.

Nowhere are the decisions that a manager must face more challenging than those made under conditions of uncertainty. From decision theory, the expected value in dollars (*E*) of a decision made under the condition of uncertainty is mathematically expressed as follows:

$$E = \rho V \qquad (3.11.1)$$

Where:

ρ : Probability of a successful outcome
V : Value in dollars of a successful outcome.

Although the expected value, E, of an outcome is increased under conditions of decreased uncertainty, the return is typically low (i.e., *V* is low) and hence the learning potential is reduced. When the value in dollars of a successful outcome (*V*) is large, there is typically a lower probability of a successful outcome (ρ). While it is true that risk averse managers will obtain a certain return on their investment, the degree of learning associated with such decisions is minimal. This can be explained by analogizing the information flow through a communication channel. The information flow through a communication channel is mathematically expressed as follows:[59]

$$I = E \, Log_2 \, (1/\rho) \qquad (3.11.2)$$

Where:

I : Information flow (bits/second)
E : Total number of events in the channel per second
ρ : Probability of a successful outcome for each event.

Equation 3.11.2 demonstrates that information flow is increased as the number of events in the channel are increased. The equation also demonstrates that the amount of information is increased as the probability of a successful outcome is decreased. Under the condition of certainty, the value of information is zero (i.e., $Log_2 \, (1) = 0$). The implications of Eq. (3.11.2) are clear. Experimentation is a cost effective imperative for gaining information under conditions of uncertainty due to the risk involved. A continuous learning organization is managed by individuals who embrace risk through experimentation. Such organizations are most likely to achieve and implement innovations that will confer the advantage(s) necessary to effectively compete and win in the marketplace. This strategy for change will be counter-intuitive to the risk averse manager who is steeped in the old logic of past strategies.

The well-known management consultant Peter Drucker[39] has often stated that efficiency is doing things right and effectiveness is doing the right things. The lack of learning inherent in maintaining the status quo is the greatest risk an organization faces. Managers should create an environment that makes it possible for individuals to experiment and innovate. It is essential for managers to understand that innovative ideas may not survive beyond their initial trials. The infeasibility of such innovative ideas represents a key learning in and of itself.

Leadership increasingly requires that organizations be created that do not define security in terms of absolute control. Loss of control or shared control is not a threat but an opportunity for reorienting the organization in a way that offers the opportunity to bring about superior levels of knowledge and performance previously not considered. Managers who litter their organizations with the stepped-on bodies of individuals who attempt to communicate or gain support for new ideas are not new. Such managers are empire builders within functional silos and are only concerned with pleasing their bosses and seldom, if ever, support anyone outside their functional group in the organizational hierarchy. They are bureaucrats that lack the necessary leadership skills for integrating horizontal communication or cooperation, and they are often preoccupied with preventing losses as opposed to improving performance. The result is an initiative-suppressing culture based on mediocrity. The inability of managers to provide individuals the opportunity to test innovative ideas adversely affects the entire work force. Management must create the organizational flexibility required to learn what they do not know.

Individuals with innovative ideas represent the visionaries of the organization. Vision is not hope. It is essential for managers to have the technical skills required to understand the risks associated with innovation. As all movement is not forward, all change is not positive. Upper level managers must find connectedness with the innovators in their organization and move forward. The central issue is not based on what the individual can do for the organization. The central issue is what the organization can do for the world through the innovator.

CHAPTER 4

Multiple Constraint
Synchronization (MCS)

4.1 Strategic Capacity Planning

Load leveling is the balancing of load with capacity. In order to determine the aggregate load capability, a capacity strategy must be established that determines the maximum rate that work can be output by the system. There are three basic capacity strategies from which a manufacturer can choose:

1. **Lead:** A proactive capacity strategy where capacity is increased based on an *anticipated* increase in demand.
2. **Lag:** A reactive strategy where capacity is increased based on a *demonstrated* increase in demand.
3. **Tracking:** A proactive or reactive strategy where capacity is increased or decreased based on an anticipated or demonstrated increase or decrease in demand. Incremental increases in capacity are relatively small as compared to the lead and lag capacity strategies.

A lead capacity strategy is potentially a high-risk strategy. If an anticipated increase in demand does not occur, the financial burden on the manufacturer is increased. The lead strategy has the advantage of improved delivery performance due to a high tolerance for disruptions such as machine breakdowns, quality problems, and unplanned worker absenteeism. The lag capacity strategy is a cost-based strategy that has the advantage of not carrying excess capacity during periods of reduced demand; however, there is a risk of poor delivery performance as a result of minimal excess capacity for responding to disruptions. The tracking capacity strategy increases or decreases capacity through the use of overtime, hiring of additional workers, or cosourcing. The tracking capacity strategy has the advantage of responding to increasing or decreasing demand patterns cost-effectively. Although the tracking capacity strategy effectively responds to short-term demand volatility, it is not an effective strategy to use under the condition of a permanent increase in demand in excess of the current available capacity level. The tracking strategy is a risk-based strategy. There is minimal excess capacity to respond to disruptions, and delivery performance may be adversely affected. It is important to note that although the tracking strategy may be effective in the short term, a sufficient time horizon should be allowed for the lead time required to augment capacity should a permanent increase in demand be anticipated. The method most often employed to augment capacity under the lead and lag strategies is structural (e.g., purchase additional technology) and infrastructural (e.g., increase the work force). The tracking capacity strategy most often employs infrastructural changes for augmenting capacity. A failure to strategically respond to permanent increases in demand in excess of the current available capacity level will damage the delivery performance of any manufacturer. The lag strategy is a do-nothing strategy that must be avoided. For high-mix, low-volume manufacturing environments, a combination of the lead and tracking capacity strategies should be used to effectively respond to demand increases or decreases.

There are three demand response capacity strategies that are of general importance when planning annual or monthly production requirements:

1. Level strategy
2. Chase strategy
3. Compromise strategy.

Consider Fig. 4.1.1. The cumulative shipments forecast exhibits periods of increased demand during periods QII and QIV that are attributable to seasonal demand shifts in the marketplace. A level response strategy is indicated by line segment α-β. The starting inventory is $10 million and the desired ending inventory is $5 million. The necessary capacity (i.e., rate of production output), R, required to successfully pursue this strategy is given by Eq. (4.1.1):

$$R = \frac{(I_e - I_b) + F}{N} \qquad (4.1.1)$$

Where:

I_e : Ending inventory
I_b : Beginning inventory
F : Forecast shipments
N : Number of periods under consideration.

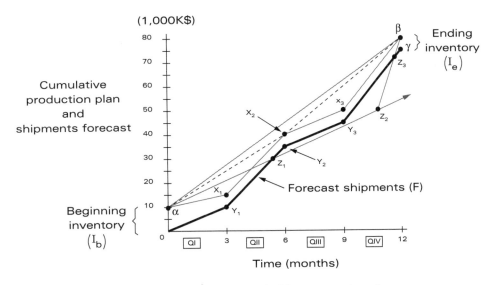

Figure 4.1.1 Cumulative production and shipments planning.

The slope of the leveling strategy line segment (α–β) represents the capacity requirement *(R)*. The inventory requirement *(I)* at any point in time *(t)* along the leveling strategy line segment can be calculated by using Eq. (4.1.2):

$$I = \left[\frac{(I_e - I_b) + F}{N} \right] t + I_b \qquad (4.1.2)$$

The level response strategy responds to forecast shipment fluctuations by using buffer inventory. Inventory levels are high during months three and nine as compared to the forecast shipments requirement. The level strategy is most appropriately used by a low-mix, high-volume manufacturer for whom cost is the competitive differentiator. In contrast to the level response strategy is the chase response strategy. The chase response strategy will minimize the inventory requirement and absorb forecast shipment fluctuations by changing capacity. The chase strategy is indicated in Fig. 4.1.1 by line segment α-x_1-x_2-x_3-β. The ending inventory of $5 million at the ends of periods QI, QII, QIII, and QIV is maintained to respond to anticipated shop floor disruptions and forecast shipment errors. For the chase response strategy, a relatively large capacity increase is required during periods QII and QIV as compared to periods QI and QIII. The result is an excess capacity condition during periods QI and QIII. When there is insufficient in-house capacity to use the chase response strategy (e.g., periods QII and QIV), the requirement for flexible capacity is addressed by using temporary workers, cosourcing,

and/or overtime. The chase strategy creates an unstable work force requirement that is less cost-effective than the level response strategy. Additionally,
excess capital equipment may have to be maintained during periods of
reduced forecast demand (e.g., periods QI and QIII). A trade-off (i.e., compromise) should be made between the high inventory cost associated with the
level strategy and the excess capacity cost associated with the chase strategy.
An arbitrary compromise response strategy is indicated in Fig. 4.1.1 by line
segment α-x_2-β. The compromise response strategy reduces the inventory
cost associated with the level response strategy and simultaneously reduces
the excess capacity cost associated with the chase response strategy. In fact,
the cost of maintaining excess capacity is minimized if, during periods QIII
and QIV, overtime alone is used to respond to increased forecast shipments.
The particular compromise response strategy chosen is a function of available
capacity. Available capacity must be equal to or greater than the maximum
required capacity for any capacity response strategy chosen. For a high-mix,
low-volume manufacturing environment, the compromise response strategy
is superior to the level and chase response strategies from a total cost standpoint, and should be adopted.

Consider the case where line segment α-z_1-y_3-z_2 represents the total available in-house capacity for a particular manufacturer. An initial capacity deficit
will occur toward the end of period QII and continue through period QIII.
The magnitude of the capacity deficit is indicated by area z_1-y_2-y_3 and may be
effectively addressed by using overtime or temporary workers. Beginning at
period QIII and continuing through period QIV, a severe capacity shortfall
will occur as indicated by area y_3-z_2-z_3. An investment in the structure and
infrastructure of the firm is most likely required to augment capacity. If this
is not possible (e.g., insufficient investment capital), cosourcing or outsourcing a portion of the manufacturing base may be the only alternatives.
Although the capacity response strategies shown graphically in Fig. 4.1.1 will
quickly indicate when and where management action is required, the production plan(s) should also be expressed numerically in order to quantify the specific actions required to correct adverse deviations.

4.2 Modeling Operations

Job shop high-mix, low-volume manufacturing environments require an
intensive focus on heuristics that are tested through simulation methods.
Serial flow, high-mix, low-volume manufacturing environments are more
amenable to analytic methodologies that can greatly assist management in the
decision-making process. Simulation is employed to derive solutions that are
too complex to solve analytically and is used as a last resort. The distinction

made with regard to simulation as compared to analysis is that simulation is trial and error. The establishment of and the rationale for a manufacturing operations model are imperative for a serial flow high-mix, low-volume manufacturer. A manufacturing operations model will greatly assist management in gaining a better understanding of problems and serve as a focal point for systematic discussion of objectives and alternatives. A manufacturing operations model should not be looked upon as an end in itself. A manufacturing operations model is only effective when it provides the means to solve manufacturing operations problems.

The modeling of complex high-mix, low-volume manufacturing environments should not be based on exaggerated claims. This will only create managerial disillusionment. It is important to note that models are useful to management as an aid in the decision-making process, not in discovering problems or implementing solutions. Additionally, models do not in and of themselves provide any assistance to management in defining goals. A high-mix, low-volume manufacturing operations model that can be regarded as an "ultimate weapon" for manufacturing decision making does not exist, and it is highly improbable that one will ever be developed.

Analytic models are almost always designed to provide specific answers to narrowly defined problems. Rarely can these results be adopted without modification if they are to be used in the broader context of differing manufacturing environments. In the process of developing a model, it is essential to identify and define those factors affecting a manufacturer's objectives. Subsequent efforts must focus on devising methods of measuring these factors and their results. The ability of a model to provide a source of standards to which results can be compared is based on the extent to which the model accurately predicts results.

With regard to the testing of new ideas (i.e., experimenting), a manufacturing operations model should indicate which experiments are worth pursuing and show how much they are worth in terms of potential profit. In essence, modeling should make explicit the implicit models of managers. The ultimate value of a model must be judged in the context of its application in the manufacturing organization, not merely in terms of its technical quality. Plainly and simply, a good model provides acceptable predictions. The ability of a model to make acceptable predictions can be judged for a period of time through experimentation, or tested based on historical data. Clearly, no model can be expected to *exactly* predict real-world manufacturing conditions.

One of the greatest dangers management faces is the situation where accidental relationships occur that are not causal (i.e., not the true cause of the effect). This phenomenon is known as *spurious correlation*. Poorly designed experiments, as well as short-term results, run the high risk of being wrong.

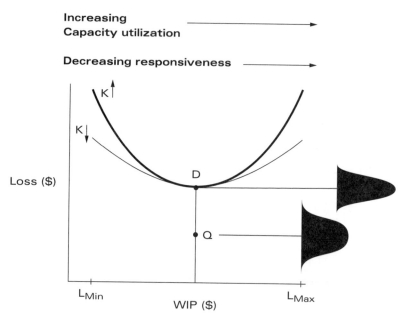

Loss ($) = K [WIP($) ± Δ]² + [Q($) + D($)]
K: sequence-independent setup cost

Figure 4.2.1 Manufacturing operations model.

If management accepts conclusions as correct that are incorrect, or conclusions are rejected when they are correct, there is a loss associated with the result. Clearly, the rejection of a correct result is the most damaging to competitiveness. Managers who preoccupy themselves with the rejection of new ideas based on fear of risk are radicals as opposed to conservatives. In the final analysis, managers must focus on the avoidance of decisions that will reject correct conclusions, and a model is an invaluable tool to use in this regard. If a model provides a basis for making better decisions than the competition, a superior model will result in a competitive advantage for the manufacturer. Thus, the model becomes a competitive weapon in and of itself. Arguing that the relationships among the many factors affecting high-mix, low-volume manufacturing environments are so ephemeral that modeling is not possible may place a high-mix, low-volume manufacturer at a self-imposed competitive disadvantage.

Consider Fig. 4.2.1. Sequencing and lot sizing methods will have a significant effect on the overall financial position of a high-mix, low-volume manufacturer. As production lot sizes are increased or decreased, there is a financial loss or benefit associated with batch errors and capacity constraint utilization as a result of the consequent increase or decrease in the *number* of batch error occurrences or sequence-independent setups performed. Financial loss varies in proportion to the lot sizing rule employed. Point Q represents the aggregate adverse expense associated with less than optimal quality levels. Aggregate adverse quality expense is comprised of defect, appraisal, and prevention costs. With regard to the horizontal axis of Fig. 4.2.1, the minimum lot size, L_{Min}, represents a lot size equal to one and the maximum lot size, L_{Max}, represents a lot size equal to the incoming order quantity. All other lot size possibilities can be represented between these two extremes using a sequencing rule for all lot size choices that is based on minimum makespan criteria. Point D represents the financial loss associated with production floor disruptions (e.g., unplanned machine downtime, unplanned worker absenteeism). The total loss function is approximated by a parabola. Based on the model of Fig. 4.2.1, it is clear that a reduction in the loss associated with adverse quality and disruptions, Q+D, will have the greatest impact on overall manufacturing performance over the entire range of lot sizes that can be produced. It is important to note that as the lot size is reduced, the resulting improvements in responsiveness and delivery performance will increase costs quadratically due to increases in idle capacity (i.e., decreased asset utilization). Note that there is also a decreased utilization of raw materials.

Although the model defines a cost minimum, it is important to understand that cost is a competitive measure. Cost plays a significant role in product pricing which, in turn, is a determinant of a manufacturer's acceptance in the marketplace. Competitive product cost and quality are marketplace "qualifiers" for a high-mix, low-volume manufacturer, while responsiveness and delivery performance are competitive differentiators (i.e., order winners). Cost is a critical factor that must be considered when establishing a particular customer response strategy. It is important to note that customers are often willing to pay a premium for superior responsiveness and delivery performance. The manufacturing model of Fig. 4.2.1 clearly indicates that a trade-off must be made between the cost of decreased asset utilization and responsiveness. Given a desired level of responsiveness and delivery performance, every effort should be made to achieve a cost position that is superior to the competition, as opposed to a cost position based on internally generated financial targets. This is most effectively accomplished through improvements in quality levels, machine preventive maintenance, sequence-independent setup cost, and proper scheduling and lot sizing techniques.

The minimum of the total loss function of Fig. 4.2.1 is established based on relationships among *internally* generated cost, responsiveness, and delivery goals. A condition is thus established where the improvement in one factor will necessarily adversely affect another factor. From economics theory, such a condition is known as a *pareto optimum*. It is interesting to note that when the minimum is based solely on internal cost factors, manufacturing operations will exhibit a degree of robustness to variability in production lead times

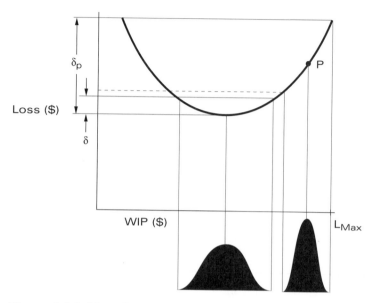

Figure 4.2.2 Manufacturing operations model with variation in WIP.

(Fig. 4.2.2). If external competitive factors necessitate an operating point (point P of Fig. 4.2.2) that is significantly shifted from the optimal operating point, the distribution of variability will have an adverse effect on production performance, and cost may increase to the extent that a manufacturer will be operating at a loss. Ultimately, the flexibility afforded to a high-mix, low-volume manufacturer is limited, based on the price the customer is willing to pay for a particular level of responsiveness and delivery performance. For this reason, it is essential to reduce sequence-independent setup times, improve quality levels, and institute an effective machine preventive maintenance program (reduce K, Q, and D of Fig. 4.2.1). The manufacturing model of Fig. 4.2.1 is an invaluable tool to use for guiding the management decision-making process.

For a high-mix, low-volume manufacturing environment, financial performance will vary significantly if lot sizes are set based on incoming order mix quantities. Consider Fig. 4.2.3. If an increase in customer orders occurs, the lot sizes will tend to increase. The increase in lot sizes will decrease the responsiveness and delivery performance of the manufacturer as well as increase the aggregate inventory investment required to support production operations (point X of Fig. 4.2.3). Conversely, if a decrease in customer orders occurs, the lot sizes will tend to decrease. The decrease in lot sizes will increase idle capacity, responsiveness, and delivery performance of the manufacturer as well as decrease asset utilization and the aggregate inventory investment required to support production operations (point Y of Fig. 4.2.3). Over time, a high-mix, low-volume manufacturer will exhibit significant

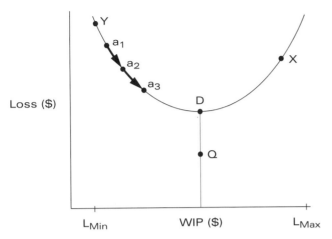

Figure 4.2.3   Manufacturing operations model with lot size = incoming
order quantity.

responsiveness, delivery, and asset utilization variability based on the dynamics of incoming order quantities. This may lead management to accept the notion that responsiveness, delivery, and inventory performance levels are a function of incoming orders. It is not unusual for manufacturers to blame poor performance on a shift in incoming order mix. If incoming orders change over a three-month period such that lot sizes shift toward the pareto optimum (points a_1, a_2, a_3 of Fig. 4.2.3), a manager may actually believe that the resulting improvement in performance represents a desirable "trend" when, in fact, this so-called trend is nothing more than spurious correlation. In certain cases, such improvements result in praise for projects designed to improve performance when they are nothing more than misguided attempts, if incoming orders continue to be the basis for setting lot sizes. Such rationale is direct evidence that manufacturing operations are being mismanaged. It is only a matter of time until manufacturing performance is adversely affected once again. For a high-mix, low-volume manufacturing environment, lot sizes *must* be based on the optimization of manufacturing objectives through the use of proper constraint scheduling *(serial flow)* or dispatching *(job shop)* techniques and cost considerations. A high-mix, low-volume manufacturer must *never* allow lot sizes to be based on incoming order quantities.

4.3 MCS Planning

The goal of MCS planning is to develop the tactical response strategy that will meet customer demand within the time horizon required by the customer at a competitive cost. A tactical response strategy is concerned with the

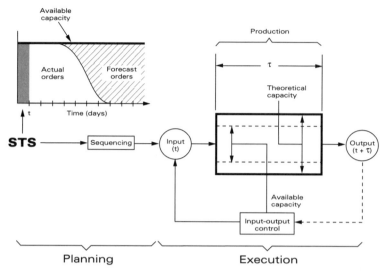

Figure 4.3.1 Manufacturing planning and execution.

detailed scheduling of operations on a daily basis. MCS scheduling for a multi-stage, serial flow, high-mix, low-volume manufacturing environment is based on the notion of transferring as much control as possible to the prerelease stage.

Consider Fig. 4.3.1. A short term schedule (STS) must be developed based on the production plan. In the short term, actual orders will normally comprise the bulk of the production requirement while forecast orders will increasingly make up the volume for scheduled periods further in the future. The STS must not violate imposed production constraints (i.e., available capacity, material availability, etc.) and the performance of the schedule according to plan must be closely monitored (i.e., input-output control).

Consider Fig. 4.3.2. Based on the entire mix of products produced, each produced product can be associated with its particular process constraint operation. The product is constrained at the particular operation where its value-adding work content time (T) is the greatest. For the case of Fig. 4.3.2, the entire mix of products offered is placed in shortest processing time (SPT) sequence under the associated process constraint operations. Additional information such as setup time (S), order quantity (Q), and forecast quantity (F) are also included for the products to be manufactured. For products that are build-to-stock or assemble-to-order, the on-hand quantity is decremented by the safety stock requirement, and the order quantity plus forecast quantity is produced only after the net on-hand quantity is exhausted. The production sequence is based on the Multiple Constraint Synchronization (MCS) algorithm.

Responsiveness for the production system is based on the total time required to negotiate *at least* one unit of each product for the entire mix of

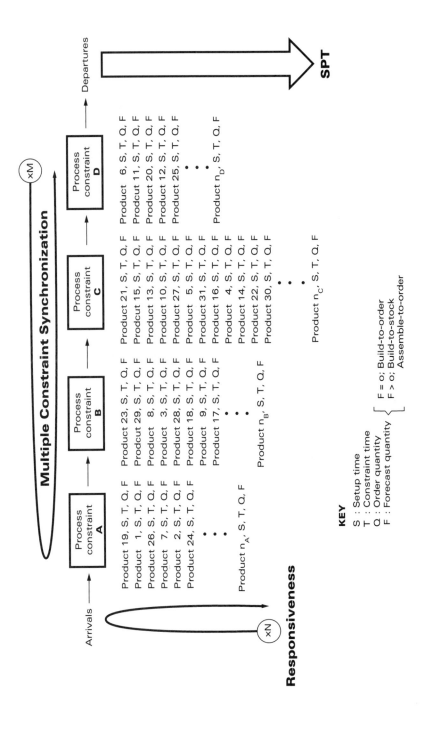

Figure 4.3.2 MCS planning tableau.

products produced. Responsiveness, lead time, WIP, and capacity utilization are functions of the lot sizing rule employed. As customer orders are received, the customer order quantity *(Q)* will increment and the forecast quantity *(F)* will decrement by an amount equal to the customer order quantity. In this manner, the forecast quantity is consumed by the incoming customer order quantity. The time horizon over which the MCS planning tableau is typically developed is one month or week, based on the production rate established during the production planning process.

Consider Fig. 4.3.3. The manner in which lot sizes are set is of particular importance. During the period of time that production operations are constrained, the processing time for all constraint operations should be as closely synchronized as possible to minimize WIP inventory. The requirement of synchronization at process constraint operations forms the basis for calculating the production lot size requirement. The minimal lot size requirement for obtaining synchronization at the process constraints is based on the notion of the system least common denominator *(SLCD)*. Frequency distributions are developed based on the total value-adding work content time associated with each constraint process step. The total monthly (or weekly) quantity for each product to be processed is multiplied by the value-adding work content time associated with each product under consideration for a particular constraint process step. Frequency distributions are also developed based on the setup time for each product associated with a particular constraint process step. Aggregate frequency distributions for value-adding work content time and setup time are developed for the overall production system, and the median value-adding work and median setup times are calculated. As a result of using the median value, the sum of the absolute values of the differences between the sets of processing time and setup time values with respect to their associated medians are minimum. Using the mean value (i.e., average value) will have a tendency to bias the SLCD calculation high or low if the distributions for processing or setup times are skewed.

The degree of amortization of setup time required is a function of the production rate developed during the production planning process. Production rates based on a compromise capacity strategy will vary for particular time periods. During periods of a decreasing capacity requirement, the number of setups will increase while lot size will decrease (e.g., periods QI and QIII of Fig. 4.1.1) while periods of an increased capacity requirement will necessitate increased lot sizes (e.g., periods QII and QIV of Fig. 4.1.1). In this manner, lot size, responsiveness, and capacity utilization will vary. Based on monthly customer order demand patterns that exhibit a "hockey stick" demand pattern, a monthly compromise strategy is developed where the production rate initially is highly responsive and successive production rate(s) are less responsive (i.e., hybrid scheduling). It is important to understand that a change in lot size

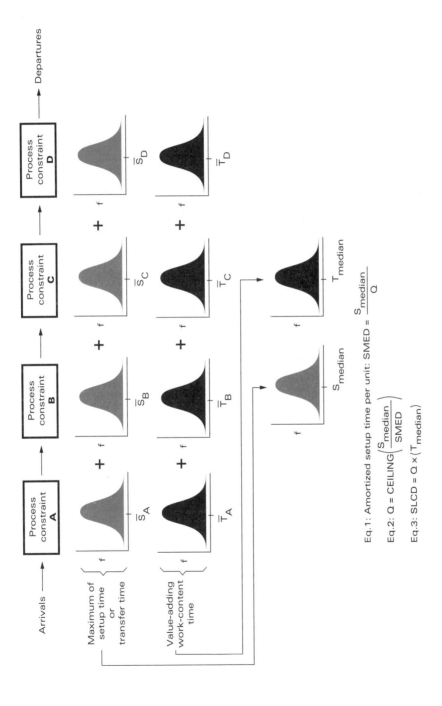

Figure 4.3.3 The manufacturing system least common denominator (SLCD).

will change capacity level within a limited range. In the case where a desired capacity increase is in excess of what can be achieved through lot size increases, it will be necessary to pursue alternative strategies such as overtime, cosourcing, outsourcing, increasing the work force, and/or purchasing additional technologies. This is also the case for the situation where increased lot sizes result in unacceptable customer responsiveness, delivery, or quality performance.

World-class setup time is often associated with the JIT concept of single-minute exchange of die (SMED)—defined as setup time that is less than ten minutes (i.e., single digit). Sequence-independent single-minute setup times (SMED) should be a goal for any high-mix, low-volume manufacturer that produces discrete products. SMED requires a manufacturing management team that is committed to continuous process improvement through total quality control (TQC) principles. Although Eq. 1 of Fig. 4.3.3 solves for the quantity required to amortize setup time to achieve SMED, the quantity required to amortize setup time should be set based on the amortization required to achieve the production rate established during the production planning process. The calculation of lot size from the amortized setup time requirement may result in a fractional quantity. For this reason, the smallest integer greater than or equal to the derived quantity (i.e., ceiling) for the amortized setup time is used (Eq. 2 of Fig. 4.3.3). The SLCD is calculated by simply multiplying the quantity required to appropriately amortize setup time by the median aggregate value-adding processing time (Eq. 3 of Fig. 4.3.3). In terms of the range of production rates that can be achieved for a given makespan, it is critical for management to understand that a company's sequence-independent setup times will confer competitive advantages in cost, responsiveness, delivery, and inventory performance relative to a competitor with greater setup times.

4.4 MCS Lot Sequencing

The development of the MCS lot sequence is based on the principle of transferring as much control as possible to the prerelease stage. Lot sequencing is based on start dates (forward-scheduling) as opposed to due dates (backward-scheduling) to ensure that production operations are not randomly sequenced. To illustrate this concept, consider Fig. 4.4.1.

The lot size *(N)* for each product is the ceiling value obtained after dividing the system least common denominator *(SLCD)* by the constraint time *(T)* for the particular product under consideration (Eq. 1 of Fig. 4.4.1). If the actual order quantity is less than the required lot size quantity, the actual

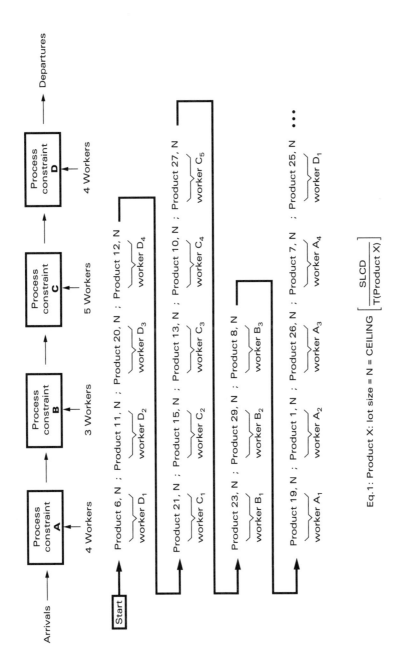

Figure 4.4.1 Lot sequencing.

order quantity plus required forecast quantity (F>0; build-to-stock, assemble-to-order) is used to satisfy the lot size requirement. If a situation occurs where the actual order quantity plus forecast quantity for a particular product under consideration is insufficient to satisfy the lot size requirement (F=0; build-to-order), the next available product in SPT sequence is used to satisfy the shortfall. A consequence of closely matching the work content times at each process constraint based on the SLCD is the resultant variability in setup time amortization that may occur. For this reason, sequence-independent setup times should be reduced as much as possible. *Setup time is a variable over which a high-mix, low-volume manufacturer must exhibit tight control.*

The lot sequencing rule is based on the Multiple Constraint Synchronization algorithm (MCS)(Fig. 4.4.1). The MCS sequence is obtained by producing the lot that is equal to the SLCD at the most downstream constraint first, followed by the lots equal to the SLCD at the subsequent constraints in upstream order (i.e., toward the gating process step). The MCS scheduled sequence for products associated with each process constraint will be iteratively produced in shortest processing time (SPT) order based on the SLCD, until a condition arises where the actual and forecast quantities associated with a particular constraint are exhausted. When this occurs, the SLCD is recalculated and the lot size requirement is updated (i.e., dynamic lot sizing). As the actual order plus forecast quantities are nearing completion, the quantity of products remaining may be insufficient to satisfy the lot size requirement. In this case, the remnants will be produced and not disrupt the overall objectives of the schedule. The MCS algorithm does not require balanced constraint times, although this is certainly desirable for WIP reduction and improved constraint utilization.

The method for establishing lot sizes presented will yield schedule sequences that are credible as opposed to optimal. Based on production rates established during the capacity planning process, balance delay calculations (Chap. 3.5) can be used to provide improved constraint synchronization. Irrespective of the lot sizing method chosen, balance delay calculations should be used to establish the *daily* quantity of particular products planned to be produced, based on available factory capacity and lot sequencing considerations. The resulting aggregate daily quantity for each product planned to be produced represents the master plan. The master plan is subsequently used to establish customer order acknowledge dates (i.e., delivery dates). In this manner, master planning and capacity planning are performed in an integrated way to achieve the company's objectives. The entire process recognizes the changes in lead time that occur based on changes in the lot size requirement. This method represents a significant improvement over the MRP II requirement to sequentially and iteratively validate the capacity requirements plan, and the MRP requirement for fixed lead times.

In the course of establishing customer order acknowledge dates, the supplier response time (SRT) must include the time to package and transport the customer order. The demand lead time (DLT) is defined as the time interval between the customer order date and the customer due date. Demand lead times are *exogenous* (definition: caused by an agent outside the system) and a supplier has minimal ability to control them. Under the condition that early deliveries are not permitted, there is an interesting relationship between the SRT and DLT. The SRT is the opposite of the DLT. The effect of an increased demand lead time on overall system performance is equivalent to a corresponding reduction in the SRT. An increase in the SRT will increase uncertainty about the future, whereas a demand lead time increase has the opposite effect. This makes intuitive sense. While the DLT is the province of marketing and sales, the SRT is controlled by manufacturing, purchasing, and distribution. It is phenomenal that such diverse and mutually exclusive activities interact as simply as they do in the course of determining the global performance of the value delivery chain. Improvements in SRT can only be achieved under the condition that lead time information is made centrally available at the point where decisions are made. Delivery date promises made for the future must not exceed the capabilities of the supplier. The customer order service function of demand management must ensure that accurate SRT and DLT information is centralized at the point where decisions are made.

The master planned quantities represent the gross requirements for the material requirements planning process (MRP). Multilevel bills of materials (BOM) should not be used to guide the MRP process (Fig. 2.6.5). Multilevel bills of materials require the stocking of partially completed products or subassemblies associated with each level of the BOM. The identification of partially completed products or subassemblies using a unique part number will not normally be required. This is true in the case where there are no service part requirements, returns (e.g., field failures), or overruns of a particular completed product or subassembly under consideration. Partially completed units are thus tied directly to planned requirements. This situation is variously referred to as phantom, transient, or blow-through assemblies. The technique used for phantom assemblies is to give them a code in the MRP data record, assume no lead time, and perform no lot sizing. The gross requirements for the partially completed product or subassembly will pass through to become the gross requirements for its associated component(s) in corresponding time periods. The code in the MRP data record is a signal that no formal work order release authorization is required. In this manner, WIP levels and system complexity will be reduced. Rather than using a work order authorization to facilitate the flow of WIP inventory, a generic Kanban signal should be used.

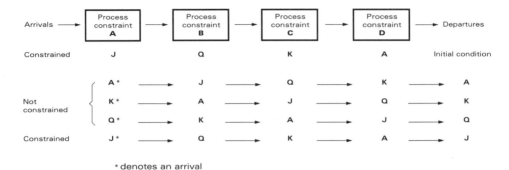

Figure 4.4.2 Balance delay.

Interprocess and intraprocess capacity imbalances associated with high-mix, low-volume manufacturing environments are inescapable. While the level of imbalance can be minimized, the residual idle capacity remaining will afford some level of buffering against processing time variability. An alternative way to grasp the concept of balance delay is shown in Fig. 4.4.2.

Production lots are symbolized as J, Q, K, and A and are constrained at process steps A, B, C, and D respectively. The initial condition is established such that all process steps are constrained. As production lots are processed, a period of time will elapse when the constraints are not constrained (i.e., balance delay). This capacity imbalance should be taken into consideration when establishing the excess capacity requirement for responding to disruptions. It is during this time interval that buffering of process time variability is afforded. In the course of executing the planned lot sequence, it is important to regulate the movement of material through the production process using the generic Kanban system of inventory control (Chap. 2.2). This will control the timing of the execution of the production sequence during production operations and minimize WIP. Additionally, transfer totes can be color coded. A unique color code would cross-reference to each process constraint and improve shop floor visibility. The sequencing of lots would be easier to control.

With the exception of repair loops, products will follow a specific sequence of operations (i.e., routing) in a serial flow, high-mix, low-volume manufacturing environment. Reentrant work resulting from repair operations should occur during the nonconstraint time period (i.e., balance delay period). If a customer requests an order change within the planning time fence, this "hot" customer order may be accommodated by supplanting a planned forecast order. In spite of these occurrences, it is important to note that the consequent predictability of processing times will continue to remain very nearly deterministic (i.e., fixed outcome). To maintain the desired level of flexibility required to respond to disruptions (e.g., machine breakdowns, unplanned absenteeism, etc.), excess capacity will have to be maintained. In the course of executing the production sequence, it is important to have real-time access to the status of open orders, machine status, customer order changes, and lot type (e.g., actual order or forecast order). In this endeavor, transactional data gathering is a must. This information, as well as feedback regarding deviations to plan, must be made available to planners and the customer order service function of demand management on a real-time basis. This is critical to achieving customer order service and inventory objectives.

An important goal of scheduling is operations control. Production operations must not become a black hole where the gating process step can be analogized as the event horizon. The requirement for information rich production operations will necessitate a strong information system. Management, planners, and production workers must have real-time access to events on the production floor. The information system should facilitate the reporting of summary information as well as detailed data and should include the following items:

- Operator availability
- Machine status
- Preventive maintenance schedule
- Lot due date status
- Lot location
- Change in order status
- Product repair status
- Engineering changes
- Action notices.

Scheduling during the planning phase is predictive while scheduling during the execution phase is reactive. Predictive scheduling is concerned with the development of schedules that facilitate accurate customer order delivery dates. Reactive scheduling involves the process of adjusting the predictive schedule when deviations between factory execution and the predictive schedule occur. The ability to respond to customer order change requests is a function of the time interval bounded by the earliest start time (i.e., forward-scheduling) and the latest completion time (i.e., backward-scheduling). If a customer order change is requested, it is essential to know what effect the desired change will have on other customer orders. The reordering of priorities (i.e., sequence), as well as other options based on real-time shop floor status information may accommodate the requested change. Customer order service and capacity utilization are the trade-offs that must continually be made in order to efficiently, effectively, and profitably satisfy customer requirements and compete in today's competitive global marketplace. The production schedule is the *definition* of how a manufacturer will accomplish this requirement.

Successful scheduling requires interactive decision support. The ability to effectively make the trade-off between customer order service and capacity utilization requires the hybrid integration of the human scheduler and the computer. Decision support scheduling systems support quantitative, mathematical, and computational reasoning as well as facilitate the freezing of changes (i.e., firm planned orders) made by planners. Real-time shop floor status information coupled with the planners' ability to understand the effect of changes to the value delivery chain, upstream and downstream, will improve the decision-making process. Rule reliant expert systems that supplant managerial judgment are an inadequate substitute for human decision making. Expert systems are a branch of applied artificial intelligence based on methods such as analogical reasoning and pattern recognition. Expert systems often result in the requirement for continuously adding increasingly sophisticated rules that are time intensive to implement (i.e., expensive). Over time, manufacturing management will most likely be at the behest of a structurally rigid, computer-based scheduling system that few will understand.

For high-mix, low-volume manufacturing environments, scheduling is a competitive imperative. The primary purpose of the MCS algorithm is to minimize makespan and thus improve throughput. When an MCS schedule is coupled with proper lot sizing techniques, competitive advantages in responsiveness, delivery, and inventory investment performance can be achieved over a less adept competitor. There are several factors that must be considered when evaluating the performance of a scheduling system:

- It should be simple to understand and use. Its simplicity is proven through its usage.
- Obtainable goals are established and the schedule provides flexibility for the unexpected.
- It provides reliable, real-time information to users. Manufacturing people can depend on its accuracy and use the information to make decisions regarding changes in the schedule in response to problems.
- Deviations in the schedule are highlighted in time for those responsible (e.g., supervisors, planners, etc.) to make needed changes. Deviations must come to the immediate attention of those responsible. Time is "always" critical.
- The schedule must be flexible enough to allow for changes to be made without disrupting the overall objectives of the schedule itself.

Companies that rise to global leadership envision a desired leadership position and establish the criteria the organization will use to chart its progress.

It can be articulated in terms of beating the competition (a.k.a. strategic intent). For a high-mix, low-volume manufacturer, competitive advantage should be pursued in quality, responsiveness, and delivery performance. Excellence in planning and scheduling is critical in this regard.

4.5 Transfer Lot Sizing

Work-in-process inventory is inherent to any multistage, serial flow, high-mix, low-volume manufacturing operation. WIP and production lead time are proportionate to the production lot size. The larger the production lot size, the longer the manufacturing lead time. Longer lead times will increase the level of WIP. The transfer lot size is the quantity of units transported to a subsequent process step from the previous process step as operations at the previous process step are completed. There is a functional relationship that can be established between the transfer lot size and production lead time. There are three basic approaches to establishing the transfer lot size:

1. *Batch* transfer lot size = production lot size
2. *Unit* transfer lot size = 1 unit
3. 1 unit ≤ *Combined* transfer lot size < production lot size.

With regard to the quantity of units transferred to a subsequent process step, the unit transfer and combined transfer lot size methods require pre-emption of the production lot size. The batch transfer lot size method requires the transference of the production lot to a subsequent process step only after operations at the preceding operation are complete. The production lead time for the batch transfer lot size method is calculated as follows:

$$\tau = Q \sum_{m=1}^{n} t_m \qquad (4.5.1)$$

Where:

τ : Production lead time
Q: Number of units per lot
t_m: Processing time per unit at the mth operation
n : Total number of operations.

The production lead time for the example of Fig. 4.5.1a is 24, *(4)(1+2+3)*. Although the number of transfers between adjacent operations is minimized, the benefit associated with reduced transfer cost is best realized when there is a significant distance between adjacent process steps. The batch transfer lot

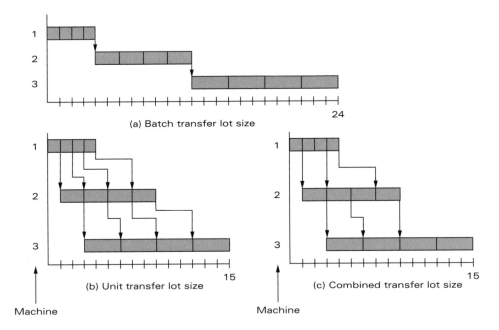

Figure 4.5.1 Transfer lot sizing.

size results in the greatest production lead time. Clearly, adjacent process steps should be placed in as close proximity as possible to facilitate the use of alternative transfer lot size methods that will reduce production lead time.

The production lead time is calculated for the unit transfer lot size method as follows:

$$\tau = \sum_{m=1}^{n} t_m + (Q - 1)t_{max} \qquad (4.5.2)$$

The production lead time for the example of Fig. 4.5.1b is 15, $(1+2+3)+(3)3$. The unit transfer lot size method yields the shortest production lead time and the number of lot transfers is maximized. This is further justification for designing a factory layout where adjacent process steps are in as close proximity to each other as possible.

In the case of the combined transfer lot size method, the transfer lot size is variable for units of production within a particularly defined production lot. The production lead time is calculated for the combined transfer lot size method as follows:

$$\tau = \sum_{m=1}^{n} t_m + (Q - 1) \left[\sum_{m=1}^{n} (t_m - t_{m-1})\beta \right] \qquad (4.5.3)$$

Where:

$\beta = 1$ if $t_m > t_{m-1}$

$\beta = 0$ if $t_m < t_{m-1}$

The production lead time for the example of Fig. 4.5.1c is 15, *(1+2+3)+(3)(1+1+1)*. Although the number of transfers is reduced for the combined transfer lot size method as compared to the unit transfer lot size method, their associated production lead times are the same. In fact, the use of either method results in the optimal production lead time. This is a special case that only holds true under the condition that the processing time per unit for the upstream process steps relative to the constraint process step (t_{max}) are in nondecreasing order (i.e., SPT), and the processing times per unit for process steps after the constraint process step are in nonincreasing order (i.e., LPT). For all other cases, the production lead times will be somewhere between the extremes of the unit transfer lot size and batch transfer lot size methods. From electrical engineering theory, the property where discrete changes are nonincreasing or nondecreasing is referred to as *monotonicity*.

The MCS scheduling algorithm develops sequences for upstream process step(s) that are in SPT sequence relative to the constraint process step, and the processing time per unit may or may not be in nonincreasing order after the constraint process step. For this reason, a combination of the unit and combined transfer lot size methods should be used. This is accomplished through the use of generic Kanban. When a generic Kanban signal is issued to a particular process step, all available processed units (one or greater than one) should be subsequently delivered to the parent process step.

It is essential to understand that production lead times are improved under the condition that the transfer lot size is less than the production lot size for any given product mix. The benefit of improved production lead time will require the design of a factory layout where adjacent process steps are in as close proximity to each other as possible. A high-mix, low-volume manufacturer that adheres to these principles will obtain competitive advantages in quality, responsiveness, delivery, and inventory performance over a less adept competitor.

At this point, it is important to note the differences between the factory layouts required for a low-mix, high-volume, repetitive demand manufacturing environment and a high-mix, low-volume, intermittent demand manufacturing environment. In addition to facilitating the transfer of small lots, the factory layout will directly affect the feasibility of using a particular Kanban technique. For the case of the single-card product specific Kanban system, a minimal number of workers is often used in a U-shaped workcell layout configuration. Produce signals, based simply on an empty stock location, will normally require a minimal number of workers to facilitate ease in determining what to produce next. The U-shaped workcell layout configuration is required to bring the gating process step in close proximity to the final process step. This is necessary to facilitate the easy transfer of the withdrawal Kanban card between the final process step and the gating process step. A linear layout configuration would render a withdrawal card transfer untenable due to the adversely long distance involved. An electronic Kanban signal could be used between the gating process step and the final process step; however, this would reduce the visibility of production flow afforded through the use of a physical card. While this logic holds true for the two-card product specific Kanban system, a significant increase in the number of workers at each process step is now possible. The produce Kanban card will establish the schedule for replenishment and as a consequence, will eliminate any confusion on the part of workers as to what needs to be produced next.

For high-mix, low-volume manufacturing environments using the generic Kanban system of inventory control, a generic Kanban signal is not passed from the final process step to the gating process step and information about what to produce is derived from a broadcast production schedule. For these reasons, the factory layout configuration for a high-mix, low-volume manufacturing environment may be either a U-shaped or linear layout configuration.

4.6 Buffer Inventory

The primary objective of buffer inventories in a multistage, serial flow, high-mix, low-volume manufacturing environment is to increase the capacity and flexibility of the production system. Buffer inventories are used to address problems associated with disruptions such as processing time variability, machine breakdowns, machine preventive maintenance, repair, unplanned absenteeism, etc.

Although the purpose for and the sizing of interprocess buffer inventories is directly related to the occurrence of production disruptions, many managers have the mistaken belief that buffer inventory is only necessary for the purpose of reconciling imbalances between supply and demand. Managerial education primarily focuses on buffer inventories that interface production operations to the external environment. The demand side of such buffers is unpredictable and forecasting techniques, cost models, and decision rules have been developed to evaluate and appropriately set such buffer inventory levels. Exogenous buffer inventory such as finished goods inventory, serves the purpose of balancing supply and demand over the long term.

It is critical for managers to understand that an imbalance between supply and demand is significantly affected by planning and scheduling decisions. Poor planning and scheduling are often the root cause of a major portion of supply and demand imbalances. Many managers use inflated FGI levels to mask the effects of poor or nonexistent planning and scheduling strategies. Interprocess buffer inventory is used to reconcile an imbalance between supply and demand within the locus of internal production disruptions that occur in the short term, as opposed to intrinsic or extrinsic external demand variability. Poor or nonexistent planning and scheduling strategies will inflate interprocess buffer inventories as well as create the need for interprocess buffer inventories that otherwise would be unnecessary. A failure on the part of management to understand and recognize the causal relationship between planning and scheduling to the supply and demand imbalance problem will perpetuate the creation of excess inventory and mortgage a company's future competitiveness in the marketplace. Buffer inventory makes explicit the implicit assumptions of managers and thus serves as a measure of managerial competence.

Multistage, serial flow, high-mix, low-volume manufacturing environments that perform inspection and repair functions do not as a general rule violate the basic assumptions of a serial flow process. This is also true for the case where parallel operations are used within a particular workcenter operation to increase available capacity.

A simple model for a production system is developed in Fig. 4.6.1a. Let us assume that the pipeline is comprised of two process steps having equal capacity and only one product will be produced at a time. Let us further consider the case where we separate the two process steps with an interprocess inventory buffer (Fig. 4.6.1b). Two dissimilar process steps separated by an inventory buffer are referred to as an *inventory bank*. In the case where the buffer is less than the maximum capacity level and production disruptions do not occur, the time required to fill the buffer (t) will increase the production lead time $(\tau_1 + \tau_2 + t)$. The increase in lead time will be equal to the time it takes to fill the buffer (Fig. 4.6.1c). For this situation, the buffer is referred to as a *series buffer*. Once the buffer is filled, the lead time of the production system is decreased to its minimum level $(\tau_1 + \tau_2)$ (Fig. 4.6.1d). The buffer is now referred to as a *shunt buffer*. If the production system continues to produce products without disruption, the buffer will serve no useful purpose and should be removed.

Let us now consider the situation where a disruption occurs at process step A. The flow of products can be sustained at process step B by consuming the buffer inventory (Fig. 4.6.2a). If a disruption subsequently occurs at process

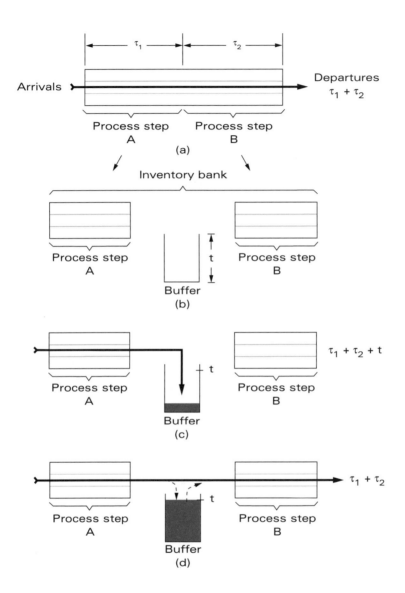

Figure 4.6.1 Interprocess buffer inventory.

step B, the flow of products through process step A will be maintained by replenishing the buffer (Fig. 4.6.2b). Based on the scenarios depicted in Fig. 4.6.2, maximum benefit is derived from the use of interprocess buffer inventory when disruptions *strictly* alternate. If disruptions occur at random intervals, the interprocess inventory buffer will be periodically empty during the time interval in which a disruption occurs at process step A. This will adversely affect the flow of products through process step B and thus reduce through-

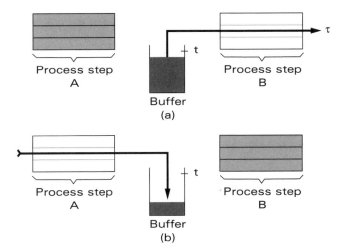

Figure 4.6.2 Interprocess buffer inventory with process step disruptions.

put. Under the condition of random disruptions, the interprocess buffer inventory level will be larger than what is required under the condition of nonrandom disruptions (e.g., scheduled maintenance).

It is important to understand the effect that unequal capacities at process step A and process step B will have on the interprocess inventory buffer. If we consider the case where the capacity at process step A is less than the capacity at process step B, the throughput of the system is controlled absolutely by the slowest process step (process step A of Fig. 4.6.3a). When the constraint operation precedes the interprocess inventory buffer, the buffer will always be empty. Alternatively, if the constraint process step follows the interprocess

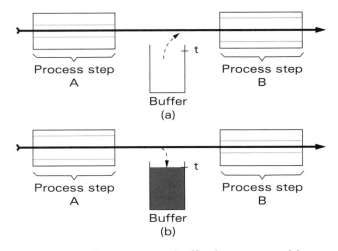

Figure 4.6.3 Interprocess buffer inventory with unequal process step capacity.

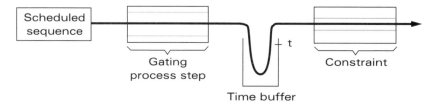

Figure 4.6.4 Theory of constraints time buffer.

inventory buffer, the buffer will always be full (Fig. 4.6.3b). Clearly, if the variability in processing time is such that the constraint operation does not shift, the interprocess buffer inventory level will be invariant (i.e., fixed).

Theory of constraints will place an interprocess capacity buffer at the input of a constraint operation in order to protect the constraint from periods of starvation (Fig. 4.6.4). Based on a production schedule, the constraint will process the contents of the buffer in scheduled sequence (i.e., series buffer). Theory of constraints refers to an interprocess capacity buffer as a time buffer. The time buffer contains that quantity of products that translate into a particular amount of processing time at the constraint. The time buffer is managed based on the average demand placed on the buffer. If the constraint empties the buffer before it is replenished, the buffer is expanded. If excess inventory is continuously present in the buffer, the size of the buffer is reduced. For a high-mix, low-volume manufacturing environment, the use of a time buffer is highly recommended under the condition of a starving constraint operation.

Analytically determining the optimal level of interprocess buffer inventory is only of academic interest. Interprocess buffer inventory levels should be closely monitored using a real-time information system. As a general rule, interprocess buffer inventories are only beneficial when there is variability in the average inventory level of the buffer. For high-mix, low-volume manufacturing environments, there is significant variability in the processing times associated with each product across the entire mix of products produced. For this reason, interprocess buffers must be managed based on the amount of processing time that products contained within the buffer represent.

CHAPTER 5

Performance Measurement

5.1 Change Paradigms

Organizational transformations are predicated by mental models that are articulated in terms of visions. Such visions are used to guide and motivate organizations toward improved levels of performance. The framework for organizational change can be categorized four ways:

1. Continuous improvement
2. Restructuring
3. Reengineering
4. Reinventing.

Restructuring is an extension of creative accounting that encompasses cost saving and cost reduction strategies. Restructuring is a short-term strategy. The driving forces of restructuring are decreasing market share or margins, and/or sluggish growth. From a cost accounting perspective, restructuring improves return on investment (ROI) performance by reducing costs through work force reduction. Although increases in net income will also improve ROI performance, the time horizon is significantly longer as compared to an asset reduction strategy. Restructuring is most often a reactive strategy employed by failed management to buy time for improving performance. Restructuring is almost always employed by corporate turnaround artists. Reducing the asset base of a company is no guarantee that improved levels of performance will occur in the future. In terms of the cost incurred when pursuing a restructuring strategy, employees are the ones who bear the cost and, more often than not, companies that focus on reducing head count fail to consider the qualifications or morale of employees who remain. The legacy of restructuring is clear: Risk has been transferred to the employee base to the extent that remuneration and employment security are contingent upon factors beyond their control. In such an environment there are fewer promotional prospects for nonexecutives. The gap that is created between management and employees is further increased when management is offered incentive-based compensation for improved ROI performance. In reality, cost cutting reduces the net worth of a company in order to achieve a higher return on investment.

A strategy-driven approach to restructuring is one in which people are redeployed in a positive way. This method focuses on process redesign where value-adding activities are maximized at the expense of non-value-adding activities. This is a cost saving strategy. Employees are thus redeployed from non-value-adding processes to value-adding processes. Competitive advantage cannot be achieved without incurring cost. The trauma of restructuring based on cost cutting will create added pressures on organizations in the long term. Such an approach is often associated with organizations that have not achieved parity with their competitors in cost, quality, responsiveness, delivery, and/or service performance. Although competitive advantages have emerged on a worldwide basis as a direct result of time-based strategies, many organizations have failed to pursue these strategies. Once parity is achieved among competitors pursuing time-based strategies, those companies that have failed to respond to this competitive reality will cease to exist. The stakeholders of a business include customers, shareholders, employees, and the local community. Cost-cutting approaches to restructuring disavow the work force and the local community as stakeholders in the organization.

Reengineering is based on the rapid and radical redesign of strategic value-adding business processes. Reengineering is initiated by starting with a clean slate. This is often referred to as a "green field" or zero-based budgeting approach to initiating radical redesign efforts. Redesign encompasses the systems, policies, and organizational structures that drive the strategic value-adding business processes. The goal of reengineering is to optimize workflow and productivity. Although functional designs of organizations are vertical, reengineering identifies and focuses intensively on cross-functional interactions to optimize the value-adding capability of the organization. Although most companies embrace change incrementally (i.e., continuous improvement), reengineering is about breakthrough change. A mindset based on the notion that change must occur incrementally is the antithesis of reengineering.

Reengineering is a socio-technological approach to process redesign that requires top management leadership. Upper management must break the chains of organizational structure in order to be successful in any reengineering effort. In fact, organizations must break out of the confining boundaries created by salary ranges and job titles. To be a contender in today's competitive marketplace, all of a company's resources must be mobilized in order to achieve radically improved levels of performance. Reengineering mandates an educated and well-trained work force. The understanding of how a company can and should be run is an essential ingredient for success in any reengineering effort. Although people are often promoted based on their knowledge of the status quo, future promotions are based on an individual's mastery of the newly reengineered organization. Plainly and simply, what once made individuals successful in the past will not make them successful in the future.

In the course of reengineering business processes, non-value-adding activities in core business processes will be eliminated. This is most often accomplished through automation. More work will be accomplished by fewer people as a result of the cost savings realized. This is in stark contrast to the cost-cutting dictum of restructuring. Reengineering seeks to attain long-term competitiveness based on a particular customer value proposition (e.g., cost, quality, delivery, responsiveness, service, technology). Reengineering is an effective method for extending known competencies, and is therefore a defensive strategy. Unfortunately, the vast majority of companies that reengineer their business processes are closing a competitive gap rather than attaining a competitive advantage.

Reinventing is about innovation and carries with it the heightened risk of uncertainty. In fact, predictability during the process of reinvention is virtually nonexistent. There is a significant difference in achieving competitive advantage as opposed to competitive parity. Competitive parity does not last for very long. More than anything else, reinventing is about "breaking the rules" for the purpose of developing core competencies previously unknown. Groups of people with new ideas are brought together to focus on innovations that will accelerate the rate of organizational learning. Such groups are often persona non grata from the viewpoint of the status quo and are often referred to as a "skunk works." Individuals within such groups are referred to as *intrapraneurs*. Upper management must break down these and other barriers to reinventing their organizations.

Winning organizations invest in innovation and are seeking to achieve advantage as opposed to simply being competitive. They fail forward. The nature of reinventing is based on the notion of accelerating failure as a learning mechanism. Communication of knowledge gained is the product of reinventing efforts. Knowing what does not work and why is just as valuable as knowing what works. Central to creating the capacity to innovate is the ability of the organization to have the flexibility to tolerate disruptions. Reinvention is not something that is "nice to do." Reinvention is mandated for those organizations that want to lead rather than follow in today's competitive global marketplace. Upper management must fund and support activities designed to facilitate the learning necessary to reinvent their organizations. Upper management must create an intensive environment of purposeful impatience (i.e., urgency). Reinvention will require individuals who work well under pressure and who can be further characterized as loving to win and hating to lose. After all, in business competition, there are only winners and losers. There is no room for incrementalist thinkers in an organization desiring to reinvent itself.

Benchmarking is the first step toward reinvention. Benchmarking is the process of identifying and understanding practices and processes from organizations anywhere in the world. The purpose of benchmarking is to avoid reinventing solutions already discovered and tested. It would be lackluster to reinvent an organization without first knowing how others have designed a similar process. An organization reinventing itself believes that it has the talent to supersede the competition on a worldwide basis. You cannot be overly humble when the goal is to be the best.

Achieving competitive advantage through reinvention is characterized by intangibles and attributes. Intangibles include the ability of the organization to collect, analyze, synthesize, and communicate information. These intangibles and the following attributes will not be found on the balance sheet and income statement, but are essential to securing competitive advantage:

- Integrity
- Passion
- Focus
- Commitment
- Competence
- Flexibility
- Timeliness
- Trust.

Leaders battle for the future while others are catching up. Contenders in the fight for the future must break away from the paradigms of the present.

5.2 Education

An organizational cultural change is mandated to create an effective learning organization. Managers must not confuse personal learning with organizational learning. Learning organizations are driven by the notion that a management team sees, hears, and listens to those things that contradict what they want to do. Thus, conflict is a competitive imperative. Although this is anathema to most managers, managers must stretch into learning domains that they cannot currently handle. Conflict avoidance is most appropriately characterized as skilled incompetence and is a primary reason why upper level managers are not always told what they need to hear in order to make effective business decisions. Nowhere is the damaging effect of conflict avoidance more pronounced than in the situation where false consensus occurs. False consensus is based on the notion that decisions are made with no dissenting voices. Individuals remain silent even though they are in disagreement. Upper level managers must test the commitment of the organization to implement new ideas. It is important to understand that you don't get the same mindset by simply declaring it. Learning experiences often occur when you get what you don't want.

If a project is worked on for several years and fails to yield the improved levels of performance originally anticipated, a manager must not get defensive and discount the fact that failure did in fact occur. Managers must tell the truth. The absolute rejection of failure is counterproductive and will only serve to disenfranchise the work force. Rapid prototyping of ideas is imperative to accelerating the learning process. If failures occur, they can be quickly recognized and learned from so that management can recalibrate and move forward.

An effective management team must not cripple the learning process by creating a "yes" organization. Organizations can evolve into a breeding ground for bad ideas by creating an atmosphere that inhibits or punishes differences of opinion. The result of such actions is called "groupthink." Groupthink is a process of rationalization that occurs when all team members begin to think alike. It is important to understand that if people are given sufficient time to think an idea through, there is an increased likelihood that weaknesses or problems will be exposed. The communication of weaknesses or problems associated with an idea can only serve to improve the learning process. Such a change will be a "cultural shock" to organizations steeped in autocratic styles of managing. Leaders must hold all major decisions up to the bright light of scrutiny in order to facilitate the learning process and, most importantly, garner commitment from the organization to implement.

Education of the work force in the fundamental principles of manufacturing planning and control is a cost-effective imperative for obtaining competitive advantage. Manufacturing disciplines such as business, marketing, engineering, and finance are staffed by highly educated professionals well equipped to perform their particular specialties. The universe of competent employees within these professions is indicated by area "U" in the Venn diagram of Fig. 5.2.1. The presence of a small subset of incompetent individuals is indicated by area "A" of the Venn diagram. The startling truth is that the proportion of all employees competent in the fundamentals of manufacturing planning and control systems is represented by area "A" while the subset of incompetent

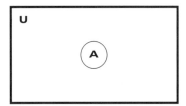

Figure 5.2.1 Venn diagram core competency assessment.

employees is represented by area "U" in the Venn diagram of Fig. 5.2.1. In fact, individuals holding degrees within the aforementioned specialties are unqualified to manage a manufacturing production environment. Fortunately, the fundamental knowledge required to know how a high-mix, low-volume manufacturing environment can and should be managed is relatively straightforward and easy to understand. Although a focus on the myriad of details can be overwhelming, the fact remains that the underlying principles are easy to grasp and must be understood by the entire organization. In fact, the underlying logic of planning and control is universal. Each and every worker should receive at least 40 hours of organization-specific manufacturing planning and control education per year. Organizational commitment will only occur when the entire work force understands "why" in addition to "what" and "how" organizational initiatives are going to be implemented.

Failure to educate the work force can result in plans that are poorly integrated. If conflict detection and resolution are performed late in the planning cycle, costly revisions and delays are likely to occur, and potential synergies are likely to be overlooked and unexploited. This will result in wasted resources. The utility of plan integration in supporting collaborative planning is best achieved if the entire work force is properly educated in the fundamentals of manufacturing. Strategic alignment, or the lack thereof, is the biggest sink of time in organizations, and paradoxically it receives the least attention. In the final analysis, customers either buy the skills, competencies, and values that a manufacturer is able to offer (strategic buy), or they buy based on competitive cost considerations (efficiency buy). In either case, a well-educated work force can provide customers with time and experience-based services that can erect effective barriers to a company's competitors such as:

- Passion for customers and customer satisfaction
- Flexibility and responsiveness
- Solutions know-how and capabilities
- Global business perspective
- Ability to manage change and align with new opportunities
- Ability to attract and retain the best people.

Leaders lead learning and make it safe rather than dangerous. Organizational learning is the continuous transformation and testing of experience into shared knowledge. Functional silos must fall. Learning is doable, fun, and required in today's competitive global marketplace. Employees of a learning organization are thoroughly engaged in what they are doing. Learning must be consciously managed. Learning requires discussion as opposed to dialog. Dialog will often allow the meaning to pass through the listener. A learning organization will transform the employee base from resources into intellectual capital assets. What emerges is an organization that is composed entirely of knowledgeable workers.

5.3 Relevance

It is critical to identify relevant measures that are most likely to advance the competitive performance of a high-mix, low-volume manufacturer, as well as predict possible financial outcomes of various management policies. Increasing customer needs versus increasing finite manufacturing resources is the dilemma that must be addressed. Manufacturers are being forced to undergo a fundamental transformation of their business performance measurement systems in an environment fraught with uncertainty and risk. The reason for this is clear. Traditional cost accounting approaches have failed to adequately lead manufacturers to improved levels of performance.

Business schools have long fostered the notion that success in managing a manufacturing business is driven by understanding the financial reports and does not require a detailed knowledge of manufacturing. This is wrong. Managing by the numbers is of value only in the context of determining whether or not a manufacturer is a viable, ongoing concern. Financial data monitors and tracks the *historical* performance of a company and does not provide so much as one scintilla of guidance about how to plan, organize, staff, direct, or control a manufacturing organization toward improved levels of performance. Business schools have been phenomenally successful in convincing their MBA students that understanding the financial reports is the key to success in managing any company. The result of managing a manufacturer by the numbers can be analogized to managing a portfolio of common stocks. A manufacturing operation is invested in or sold based on the numbers. Implicitly assuming that the performance of the past is a predictor of the future by managers with minimal or no education in manufacturing planning and control systems will most likely result in the divestiture of a manufacturing operation that is performing poorly. There is a big difference in knowing how a manufacturing company is managed, versus knowing how a manufacturing company can and should be managed.

Business school graduates who have been indoctrinated in the concept of management by objectives (MBO) are taught that performance is most appropriately measured by the following criteria:

- Quality
- Customer service
- Inventory turnover
- Profit margin
- Labor efficiency
- Machine utilization
- Budgeted overhead cost.

What they are not taught is that it is *impossible* to independently achieve improvements in all of these measures. Any attempt to improve performance in one measure will result in the degradation of another measure. The result is a management team that is entrapped in a never-ending cycle of chasing suboptimal performance measures, never understanding the fundamental relationships causing this calamity. Plainly and simply, a manufacturing company must be managed as an integrated whole.

Relevant measures are based on time and physical units of measure. Specifically, improved levels of performance are directly related to the flow of materials and information. Continuous improvement in the flow of materials and information is the prescription for obtaining a competitive advantage in quality, cost, delivery, responsiveness, and service. The notion of time as it relates to the flow of materials and information is often referred to as "Theory-T," where "T" is time.

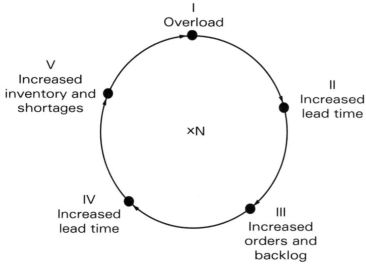

Figure 5.3.1 Lead time malady.

Fundamental to the proper management of time in any manufacturing company is the requirement to not overload production operations. The overloading of production operations will result in certain disaster and is one of the primary contributors to a management decision to divest their manufacturing base. The disastrous consequences of overloaded production operations result in what is referred to as the lead-time malady, depicted in Fig. 5.3.1. If customer orders are promised for time periods when insufficient capacity exists *(I)*, a manufacturer may inappropriately increase quoted customer lead times *(II)*. Customers will subsequently respond to increased lead times by increasing demand and customer order service will generate more factory and/or purchase orders *(III)*. Increased factory and/or purchase orders will further increase lead time and backlog *(IV)*. Increased lead time will cause a proportionate increase in uncertainty of the master plan for periods extended further into the future. Customer order change requests will vary significantly as the priority (i.e., delivery date) for their order gets closer. If the customer order is decreased, the manufacturer will be forced to carry excess inventory. If the order is increased, the lead time will increase based on the time to acquire additional material. Priorities for all customer orders will become extremely hard to manage and material shortages will increase. Stockouts due to shortages arising from customer order changes will cause purchasing to inflate safety stock levels to buffer against the increased uncertainty associated with increased lead times *(V)*. The further overloading of production operations will subsequently cause the manufacturer to further inflate lead times, and the cycle will repeat. As lead times grow out of control, excess inventory, stockouts, and excess backlog will also increase. Amazingly enough, the manufacturer may begin to actually believe that growth in business is occurring and invest in the structure and infrastructure necessary to respond to this mirage. Once the available capacity level is increased to the point where the production rate is in equilibrium with "apparent" demand, further lead time increases will cease. Customer order demand volatility will now exhibit periods of decreased demand and create an excess capacity condition. In order to alleviate the resulting excess capacity condition, lead times will be reduced in order to attract more customers. Reduced lead times will decrease uncertainty about the future and customers will alter their demand quantities to align with decreased lead time. As customer order demand quantities are reduced, excess capacity will begin to grow to the point that the manufacturer will believe that a business downturn or recession is occurring. Lead time decreases will continue until the cost of idle capacity is to the point that the manufacturer is no longer competitive. In fact, many companies have been driven to bankruptcy or have gone out of business as a result of falling victim to this vicious cycle. The semiconductor and machine tool industries are cases in point.

Top management must demand qualified professionals who thoroughly understand the strengths and weaknesses of the many manufacturing planning and control techniques available, as well as how to implement them. The holding of the American Production and Inventory Control Society (APICS) Certification in Production and Inventory Management (CPIM) should be a necessary condition for holding managerial rank in this field.

In the course of optimizing the flow of material and information, it is essential to structure performance measures around the planning and execution components of a manufacturing operation. High-performance execution is predicated by excellence in planning and data integrity. Fast feedback is necessary, and is only effective when coupled with sound analysis and judgment which, in turn, ensure efficient and effective corrective action.

Improvements in lead time, inventory, and capacity are most appropriately effected by a relentless search for and elimination of waste. Improvements are most effectively supported by performance measurements based on the maximization of the ratio of value-adding to non-value-adding time. Waste represents the greatest barrier to improved manufacturing performance and the following list is not exhaustive:

- Defects
- Overproduction
- Unnecessary operations
- Excess inventory
- Unnecessary human effort
- Delays
- Material transportation
- Unnecessary meetings.

An often overlooked source of waste is the excessive time required to make decisions and act on them. Although the need for making better decisions is often recognized, very few have devoted any energy to systematically examine how decisions are made. Disjointed or time intensive decision-making processes require a systematic evaluation in order to determine if a particular decision can be eliminated altogether through process simplification or passed to lower levels in the organization.

Clearly, lead time, inventory, or capacity buffers are ineffectual opiates for improperly managing a manufacturing company. Fundamental to the effective control and improvement of lead time, inventory, and capacity is the elimination of waste, and the balancing of inputs with outputs based on proper lot sizing and scheduling techniques.

Nonfinancial rather than financial measures yield the most relevant measures of performance for any manufacturing environment. Nonfinancial measures are used to track performance to goals that are challenges to be continuously improved upon, and they must align in an integrative way from the lowest to the highest levels of the organization. Only in this manner will the financial performance of the company continuously improve. *Financial performance measures record history while nonfinancial measures make history.*

5.4 Metrics

The term metrics is used to denote performance measures at the highest level of a business. They are the measures that meet the needs of stakeholders and the strategic goals of the organization. Central to metrics are the measures of revenue, profit, operating cost, and capital investment. As required by law, these measures are reported externally via the balance sheet and income statement. Metrics are used to establish the top level goals for organizational performance and are measures against which all other competing alternatives are compared. Metrics focus on the sales, marketing, research and development, and manufacturing segments of a business. Fundamental to the improvement of financial performance metrics is the measurement and improvement of the following:

- Sales
- Product performance
- Manufacturing capacity
- Manufacturing productivity.

Global measures (i.e., metrics) should represent a balanced approach toward guiding operating behavior as well as strategy setting. The global drivers of manufacturing performance are most appropriately categorized as follows:

- Throughput
- Inventory
- Operating expense.

Throughput is defined as total revenue generated through sales. Throughput is not a measure of the total revenue associated with products produced. Thus, finished goods inventory is not included in the measure of throughput. Inventory is the total financial investment in materials intended to be sold, and operating expense is the total financial investment required to convert the inventory investment into throughput. Inventory carrying cost is included as an operating expense. Throughput, inventory, and operating expense are measured and reported for specific time periods (e.g., monthly).

When considering the measurement of throughput, the financial loss incurred by a manufacturer due to disruptions (e.g., defects, machine breakdowns, etc.) at constraint or bottleneck operations is based on the selling price of the product(s) being disrupted. Although velocity costing is a viable means to cost a product for the purpose of establishing a selling price, it is not an appropriate tool to use for measuring financial loss due to disruptions at constraint or bottleneck operations. Velocity costing will cost products differently at each process step based on time considerations. The financial loss associated with disruptions at constraint and bottleneck operations is based on the sales price of the particular products affected and not the financial loss (i.e., cost) associated with a particular process step.

5.5 Inventory Turnover

Inventory turnover is one of the least understood of all performance measurements. The efficacy of inventory turnover as an effective tool for managing a company's inventory investment is significantly increased when coupled with a proper understanding of its dynamic behavior. The standard measure of inventory turnover is the ratio of manufacturing cost of goods sold (MCGS) (i.e., the cost of material, overhead, and labor) to the *average* inventory investment. An alternative method of calculating inventory turnover based on *transitory* inventory rather than average inventory will result in a condition where inventory turnover responds almost instantaneously to changes in business conditions. Transitory inventory is defined as the aggregate

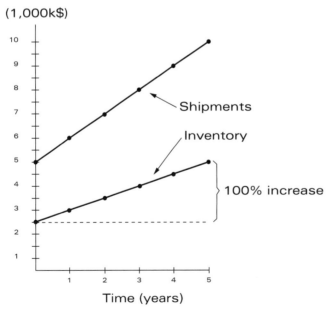

Figure 5.5.1 Shipments and inventory versus time.

inventory investment at any point in time. When calculated correctly (i.e., using transitory inventory), inventory turnover is a time sensitive measure. Changes in inventory turnover can represent significant changes in inventory investment levels.

Consider Fig. 5.5.1. If the total inventory investment is considered in isolation, a 100 percent increase in inventory investment will most certainly draw attention. However, if we note that the shipments to inventory ratio is fixed at two, we can conclude that the inventory turnover ratio indicates that the increase in inventory can be attributed to an increase in sales.

The valuation of MCGS and total inventory must be based on standard cost. Standard cost is a targeted cost whereas direct cost (i.e., variable cost) changes in proportion to volume increases and decreases. Standard cost will establish a baseline for calculating inventory turnover. MCGS must include total shipments and not include customer orders that are promised for the future (i.e., bookings).

Although customer orders are highly volatile, shipments will be relatively stable if load is balanced with available capacity and proper scheduling and lot sizing techniques are employed. Bookings do not represent the actual productive output that current inventory investment levels support and must not be included in the total shipments calculation. Although nontrade purchases are not normally included in the total shipments calculation, intercorporate, subsidiary, and field service replacement parts must be included in the total shipments calculation.

The total inventory investment is calculated based on standard cost. The total inventory investment will include all of the following types of inventories and this list may not be exhaustive:

- Raw material
- Work-in-process
- Finished goods
- Pipeline (i.e., in-transit)
- Defective
- Repair material
- Obsolete material
- Consigned material
- Scrap.

It is important to understand that the aggregate inventory investment is based on the type of manufacturing environment employed (e.g., build-to-stock, assemble-to-order, build-to-order) as well as considerations unique to a particular manufacturer (e.g., manual or automated processes). Based on such considerations, the objectives for inventory turnover can be appropriately established. In fact, inventory turnover objectives can be calculated based on raw material, work-in-process, and finished goods inventory levels. A detailed analysis of these categories of inventory will facilitate the uncovering of the non-value-adding components of inventory investment. Clearly, reductions in the non-value-adding components of the total inventory investment will result in a consequent improvement in inventory turnover performance.

Inventory turnover should not be calculated as the ratio of MCGS to average inventory investment. Average inventory will comprise past time periods and inventory turnover will only serve as a historical performance measure. This will put management in the reactive mode when adverse deviations occur. A time lag in transitory shifts in the aggregate inventory investment will often result in a misrepresentation of dynamic changes in inventory due to the smoothing affect of averaging. Using average inventory as a mechanism to calculate inventory turnover will seriously hamper a manufacturing organization's ability to institute corrective actions when adverse deviations occur. This can damage a manufacturer's competitiveness in the marketplace.

The total MCGS is based on actual shipments, excluding forecast shipments planned for the future. The total annualized MCGS will therefore have to be extrapolated based on current sales levels or the average of sales levels associated with the most recent past such as the prior three months. By annualizing MCGS based on the previous three months' sales performance, a degree of stability will result as compared to annualizing MCGS based on the current month's sales performance. This is true under the condition where poor scheduling and lot sizing techniques are being employed. When proper scheduling and lot sizing techniques are employed, shipment revenues and hence MCGS will be relatively stable on a monthly basis. This will facilitate the annualizing of MCGS based on the current month's shipments' level, and further increase the sensitivity of the inventory turnover measure of manufacturing performance. Additionally, the measurement of MCGS and total inventory investment can be based on time (e.g., weeks of supply). The measure of inventory turnover is given by Eq. (5.5.1):

$$Inventory\ Turnover = \frac{\sum(3\ month:\ MCGS)\times 4}{Transitory\ Inventory}\ or\ \frac{\sum(1\ month:\ MCGS)\times 12}{Transitory\ Inventory}$$

$$(5.5.1)$$

Consider the case of a high-mix, low-volume manufacturing environment where MCGS is $600 million per year. A decrease in inventory turnover from 4.6 to 4.5 turns will yield a $2.9 million increase in inventory. This increase will most likely draw the attention of management. In order to avoid unnecessary overreactions on the part of management to inventory turnover deviations and facilitate fast response time to significant deviations, the resulting change in inventory level that occurs due to a change in inventory turnover should be toleranced. It is also important to understand that a significant increase in inventory turnover may also indicate that significant manufacturing problems are occurring. A significant increase in inventory turnover may be indicative of stockouts. This will severely damage the ability of a manufacturer to meet promised customer delivery dates.

Inventory turnover must be a dynamic rather than historical (i.e., average) measure of inventory performance in order to increase its utility as an effective inventory management control tool. The proper understanding of and use of the inventory turnover performance measure will increase the ability of a manufacturer to more effectively utilize its assets and obtain a competitive advantage over less adept competitors.

5.6 Productivity

Productivity is defined and expressed as the ratio of salable output produced (i.e., throughput) to input resources consumed by a given organizational unit. Productivity measurement provides insight into how effectively a manufacturer is utilizing its constraint or bottleneck resources. The utility of productivity measurement as a management control tool is increased when changes in productivity are monitored over time. Productivity measurement in future time periods is compared to a baseline measure of productivity. The baseline time period is usually based on a one-year time horizon in order to fully capture a representative sample of a manufacturer's production history. For obvious reasons, a baseline year should not have exhibited extreme abnormalities or problems. This will only distort the baseline measure of productivity. The current period measure of productivity is compared to the baseline productivity measure in order to calculate the change in productivity performance that occurred. While baseline productivity is measured for a one-year period, current period productivity is measured over a time period less than or equal to the baseline time period. The general equation for calculating a percentage change in productivity (ΔP) is given by Eq. (5.6.1):

$$\Delta P = \left(\frac{P_C}{P_B}\right) 100 - 100 \qquad (5.6.1)$$

Where:

$$P_C = \frac{\textit{Current period output}}{\textit{Current period input}} \quad \textit{and} \quad P_B = \frac{\textit{Baseline period output}}{\textit{Baseline period input}}$$

The baseline or current period input is calculated as the sum of labor, material, and capital. These partial input terms should be expressed in common values such as equivalent baseline year dollars. Partial productivity measures are expressed as the ratio of the baseline or current period output to a single input resource (i.e., labor, material, or capital). Partial productivity measures disaggregate the total productivity measurement into its component parts, which provide insight into the cause of change in the total productivity measurement. Output is measured monetarily based on throughput. Labor should be measured based on total cash compensation, including the cost of fringe benefits. Material is measured monetarily, and capital input is measured based on equipment hours.

It is important to note that the partial productivity measure based on capital input (i.e., automation) is a dangerous measure if productivity improvement is based on standard cost accounting measures. A standard cost accounting approach will attempt to maximize the utilization of all capital resources in order to obtain a return on the book value of capital investment in the shortest possible time horizon. Inferences about the productive value of a capital resource or its economic life cannot, and should not, be made based on the book value of the capital resource. The maximization of utilization of a non-constraint resource (labor or capital) is counterproductive. Additionally, the maximum utilization of a constraint or bottleneck resource is limited by the percentage balance delay established during the production planning process. The production rate established during the production planning process determines the utilization level for labor and capital resources. It is essential to understand that excess capacity is required to afford a manufacturer the flexibility to respond to disruptions and customer order change requests. Clearly, the maximization of utilization of a capital resource will not serve this purpose. Productivity measurement must be understood and used carefully. Cost accounting approaches to improving productivity performance will result in decisions that will damage a manufacturer's competitiveness in the marketplace. *Cost accounting approaches to optimizing machine utilization rates are irrelevant and must not be used.*

Fundamental to the concept of productivity is efficiency. Consider Fig. 5.6.1. Although productivity measurement is based on value-adding input, productivity improvements are most effectively achieved by focusing on and eliminating the non-value-adding components of manufacturing. Productivity improvement is therefore most effectively achieved by focusing on improvements in efficiency. Continuous quality improvement, improved preventive

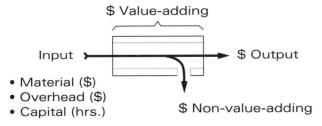

Figure 5.6.1 Efficiency.

maintenance, reduced sequence-independent setup times, and proper scheduling and lot size rules are the central foci for reducing manufacturing lead time and improving efficiency.

A manufacturing organization must break out of the mode of readdressing recurring production problems. This is accomplished by identifying and eliminating the root causes of production problems through common sense and scientific methods. Efficiency is most appropriately expressed as a function of the non-value-adding component of production lead time. Efficiency is an "output side" measurement, and is calculated as shown by Eq. (5.6.2):

$$Efficiency = \frac{\$Value\text{-}adding}{\$Value\text{-}adding + \$non\text{-}value\text{-}adding} \qquad (5.6.2)$$

The measurement of manufacturing operations effectiveness is calculated as shown by Eq. (5.6.3):

$$Effectiveness = \frac{\$Actual\ Output}{\$Expected\ Output} \qquad (5.6.3)$$

Effectiveness is a function of efficiency. Productivity can be expressed as a function of effectiveness and efficiency, and is calculated as shown by Eq. (5.6.4):

$$Productivity = \frac{\$Actual\ Output}{\$Value\text{-}adding + \$non\text{-}value\text{-}adding} \qquad (5.6.4)$$

It is important to note that productivity can be functionally related to the calculation of profit. Profit is typically calculated as shown by Eq. (5.6.5):

$$\$Profit = \$Revenue - \$Cost \qquad (5.6.5)$$

An equally valid method for calculating profit is shown by Eq. (5.6.6):

$$\$Profit = \$Revenue\left(1 - \frac{1}{Productivity}\right) \qquad (5.6.6)$$

Equation (5.6.6) is an important result. Increased profits occur under the condition of stable productivity and increasing revenue. Increased profits will also occur under the condition of stable revenue and increasing productivity. The revenue-cost relationship of Eq. (5.6.5) will misguide management toward a focus on cost cutting to increase profit. Restructuring efforts will subsequently focus on the reduction of head count that reduces the value-adding component of manufacturing production cost. Equation (5.6.6) will focus management on the non-value-adding component of manufacturing cost that will improve the manufacturer's competitive position in the market-place.

A structured approach must be taken toward maximizing the ratio of value-adding to non-value-adding production operations cost. When optimizing a manufacturer's value-adding capability, the reduction of non-value-adding activities should occur prior to investing in new technologies and control systems. This is most effectively accomplished through a relentless search for and elimination of waste. The most significant waste for many manufacturers is that of idle inventory (e.g., warehouse, store, interprocess buffer, and finished goods inventories).

A total analysis of the value delivery chain must be performed by calculating the value-adding component of production lead time and the inventory cost associated with idle inventory. The value-adding time that inventory spends in the production system is typically a small fraction of the non-value-adding (i.e., idle) time that inventory spends in the production system. It is important to understand that effective scheduling and lot sizing techniques must be employed to leverage the improvement in manufacturing efficiency resulting from the reduction of non-value-adding activities such as reduced sequence-independent setup time, move time, machine down time, repair time, etc. The conversion efficiency of a manufacturing production operation can be modeled as shown in Fig. 5.6.2.[9]

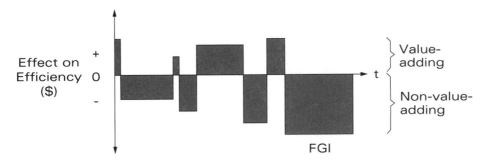

Figure 5.6.2 Effect on conversion efficiency versus time.

Velocity costing is used to quantify the associated cost at each process for the time inventory spends as non-value-adding. A positive conversion cost represents a value-adding component of manufacturing production cost while a negative conversion cost represents a non-value-adding component of manufacturing production cost. The inability of a manufacturer to determine the value-adding and non-value-adding components of production lead time is indicative of a manufacturing production environment that is operating out of control.

The time that inventory spends in the production system as value-adding is so small that it is astonishing that management focuses as much attention on capital investments in machine technology and information systems as it does today. A focus on and investment in higher performance machine technology and information systems will only serve to improve the value-adding component of manufacturing lead time. Without a concerted effort focused on the reduction of the non-value-adding component of manufacturing lead time, the purchase and installation of an improved information system will only serve the purpose of enabling managers to make poor decisions more rapidly. The new information system will thus create more problems than it will solve.

A focus on the value-adding component of manufacturing lead time, coupled with a standard cost accounting approach to recovering the book value of new investments in technology, can lead to disaster. Work-in-process levels may increase to the extent that the carrying cost of excess inventory is far in excess of revenue generated through sales. When reductions in the non-value-adding component of manufacturing lead time are coupled with proper scheduling and lot sizing techniques, a manufacturer can achieve advantages in quality, cost, delivery, flexibility, and responsiveness over a less adept competitor.

5.7 Operations Performance

The symptoms of, and rationalizations for, manufacturing operations financial performance failures are numerous:

- Inaccurate forecasts
- Poor inventory record accuracy
- Poor engineering data (e.g., bills of material)
- Unreliable vendors
- Lack of management commitment.

Although these symptoms are often associated with the cause of manufacturing operations performance failure, they stem from a lack of understanding on the part of management about how to create systems and measures consistent with the strategic and organizational requirements of the company. It must be understood that the performance of any manufacturer is based entirely on the performance of its systems and processes. Manufacturing operations performance measures are a must, and are often referred to as process performance measures (PPMs). The purpose of any performance measurement system is to encourage a positive attainment of goals. As a general rule, people always attempt to improve performance in a manner based on how they are measured. There must be clear lines of communication to all levels of the organization about which financial and nonfinancial performance measures are used, and actual levels of performance must be open to exposure.

A plethora of on-floor priority decisions ultimately determine the output of a factory. The manner in which priorities are set is a significant determinant of overall performance characteristics. Priority considerations are linked to system design considerations that provide linkage to strategies formulated at the appropriate organizational level. In this regard, top management must supply more than moral support. Top management must be involved in setting the strategy of the organization. With regard to strategy, the architecture of the organization defines the extent to which decentralization occurs vertically. Strategic focus determines the extent of decentralization horizontally. Organizational alignment with the strategic intent of a manufacturer will guide general responsibility assignment patterns for all production employees as well as the relationship of these patterns to systems and performance measures. It is important to understand that decision-making patterns cannot support the manufacturing strategy when they are limited in scope or made on a stand-alone basis. It is the interrelated set of decision-making patterns that facilitates the efficient and effective achievement of a manufacturing strategy. All manufacturing decisions are related in important ways. The central issue is the assignment of responsibilities for ensuring the integrity of the control system.

Functionally based organizational architectures will not and cannot ensure success, and integration within a vertically aligned functional architecture does not help. The achievement of performance measurement goals mandates a cross-functionally integrated organizational architecture. Only in this manner can a manufacturer achieve strategic competitive advantage. Functions must not act in a manner that is limited by the scope of their specialization. *Achieving a strategic competitive advantage is much less proactive and reactive than it is interactive.*

Table 5.7.1 Performance measures

PERFORMANCE MEASURE (%)	METHOD OF CALCULATION
Business plan	$\left(\dfrac{Actual\ ROI}{Planned\ ROI}\right)100$
Sales plan	$\left(\dfrac{Orders\ Booked}{Planned\ Sales}\right)100$
Production plan	$\left(\dfrac{Actual\ Production}{Planned\ Production}\right)100$
Master production schedule	$\left(\dfrac{Actual\ MPS}{Planned\ MPS}\right)100$
Material requirements plan	$\left(\dfrac{Orders\ With\ Accurate\ Due\ Dates}{Total\ Open\ Orders}\right)100$
Capacity plan	$\left(\dfrac{Hours\ Released}{Hours\ Planned}\right)100$
Bills of material accuracy	$\left(\dfrac{Correct\ Bills}{Total\ Bills\ Audited}\right)100$
Inventory accuracy	$\left(\dfrac{Correct\ Parts}{Total\ Parts\ Counted}\right)100$
Routing accuracy	$\left(\dfrac{Correct\ Routings}{Total\ Routings\ Audited}\right)100$
Purchasing	$\left(\dfrac{Items\ Received}{Items\ Due}\right)100$
Production control	$\left(\dfrac{Operations\ Complete}{Operations\ Due}\right)100$
Delivery	$\left(\dfrac{Orders\ Shipped\ On\ Time}{Total\ Orders\ Shipped}\right)100$

Table 5.7.2

CLASS	TOLERANCE (%)	CHARACTERISTICS
A	95	Complete closed-loop system. Top management uses the formal system to run the business. All measures average 95 to 100 percent.
B	80	Formal system in place but not working effectively. Top management approves plans without participation. Measures average 80 to 95 percent.
C	70	MRP is order-dispatching. Formal and informal systems run the business. Subsystems are missing. Measures average 70 to 80 percent.
D	50	Formal system is working poorly or is not in place. Minimal or no management involvement. Minimal or no user confidence. Measures average 50% or less.

Controllers (a.k.a. comptrollers) are responsible for financial metrics, and financial analysts are responsible for financial subgoals that are in alignment with metrics. Although financial goals are historical measures, nonfinancial manufacturing performance measurements can be directly traced to financial results in all dimensions. Macro nonfinancial measures begin at the business planning process and continue through factory floor execution. The *minimum* set of nonfinancial measures required to establish the baseline for quantifying overall MRP II system performance is indicated in Table 5.7.1. All measures are expressed as percentages.

It should now be apparent that all critical nonfinancial performance measurements are expressed as ratios of actuals to planned levels (i.e., goals) of performance. In the course of monitoring nonfinancial measures, it is critical to evaluate performance levels based on tolerances in order to account for random performance variations that occur over time. Tolerances are primarily useful as a flag to indicate whether or not management intervention is required. Additionally, performance measurement deviations can be used to establish a benchmark for MRP II performance. The benchmark classification of a manufacturer using an MRP II planning and control system is indicated in Table 5.7.2.

The goal for any company using an MRP II planning and control system should be Class A MRP II performance. A Class A MRP II level of performance is indicative of a highly educated and disciplined work force that demonstrates predictability in the achievement of planned levels of performance on a consistent basis. For a high-mix, low-volume manufacturer, Class A MRP II performance is an essential stepping stone toward world-class levels of performance.

5.8 World Class

World-class manufacturers create systems and processes that facilitate superior levels of flow in both information and products. Such organizations are characterized by management teams that have coherence of vision, consistency of implementation, strength of leadership, and a healthy fear of organizational complacency. World-class manufacturers exhibit the following four characteristics:

- The preferred supplier of products, information, and services in world markets
- The preferred choice of capital investment in world financial markets
- The primary choice for employment
- The preferred business choice of the community.

World-class manufacturers have established cultures where continuous learning, continuous improvement, and reinventing through experimentation are a way of life. Detailed nonfinancial performance measurements are used to chart their progress toward achieving competitive advantage in key success factors such as cost, quality, delivery, flexibility, responsiveness, service, and technology. The following examples of world-class performance measures are expressed as percentages, and this list is not exhaustive:

1. Reduction in setup time
2. Reduction in lot size
3. Reduction in cycle time
4. Reduction in floor space
5. Reduction in engineering changes
6. Reduction in warranty cost
7. Reduction in total cost of quality
8. Increase in on-time delivery
9. Increase in output per employee
10. Increase in inventory turnover
11. Increase in revenue based on new products
12. Increase in common parts per product
13. Increase in processes mastered per employee
14. Increase in internal customer satisfaction levels (survey)
15. Increase in external customer satisfaction levels (survey).

World-class manufacturers do not come into existence by declaring their desire to become world-class manufacturers. World-class manufacturers learn from their mistakes and achieve continuous improvements in all aspects of the business enterprise through a combination of evolutionary and revolutionary changes. Ultimately, world-class manufacturers come into existence based on their ability to tap the potential of their work forces through a combination of employee organizational education and involvement. How about you?

Bibliography

1. Adams, J. L., *The Care and Feeding of Ideas*, Addison-Wesley, 1986.
2. Aggarwal, S., "Flexibility Management: The Ultimate Strategy," *Industrial Management*, November–December 1995.
3. Agin, N., "Optimum Seeking With Branch and Bound," *Management Science*, vol. 13, no. 4, December 1966.
4. Anderson, D. M., *Design for Manufacturability*, CIM Press, 1990.
5. Ashour, S., "An Experimental Investigation and Comparative Evaluation of Flow Shop Scheduling Techniques," *Operations Research*, vol. 18, no. 3, May 1970.
6. Awasthi, V. N., "ABC's of Activity-Based Costing," *Industrial Management*, July–August 1994.
7. Baker, C. T., and B. P. Dzielinski, "Simulation of a Simplified Job Shop," *Management Science*, vol. 6, no. 3, April 1960.
8. Baker, K. R., *Elements of Sequencing and Scheduling*, Dartmouth College, 1995.
9. Barker, R. C., "Restructuring Production Operations: The Application of Time-Based Value-Adding Models," *International Journal of Advanced Manufacturing Technology*, 1992, pp. 225–230.
10. Baumol, W. J., *Economic Theory and Operations Analysis*, Prentice-Hall, 1972.
11. Bennett, M., and J. H. F. Sawyer, "The Uses of a Simulation Model in Studying a Sequencing Problem in a Batch/Flow Environment," *International Journal of Production Research*, vol. 9, no. 4, 1971.
12. Benson, A., "Assembly Machines and Systems: Focus on Flexibility," *Assembly*, vol. 38, no. 7, July–August, 1995.
13. Blackburn, J. D., *Time-Based Competition*, Irwin, 1991.
14. Blackstone, J. H., *Capacity Management*, South-Western, 1989.
15. Blumberg, D. F., *Managing Service as a Strategic Profit Center*, McGraw-Hill, 1991.
16. Breightler, C. S., D. T. Phillips and D. J. Wilde, *Foundations of Optimization*, Prentice-Hall, 1979.
17. Bronfenbrenner, M., W. Sichel, and W. Gardner, *Macroeconomics*, Houghton Mifflin, 1987.
18. Brooks, G. H., and C. R. White, "An Algorithm for Finding Optimal or Near Optimal Solutions to the Production Scheduling Problem," *Journal of Industrial Engineering*, vol. 16, no. 1, January 1965.
19. Brooks, R. B., and L. W. Wilson, *Inventory Record Accuracy*, John Wiley and Sons, 1995.
20. Buffa, E. S., *Meeting the Competitive Challenge*, Irwin, 1984.

21. Bunn D., *Analysis for Optimal Decisions*, John Wiley and Sons, 1982.
22. Buzacott, J. A., "The Role of Inventory Banks in Flow-Line Production Systems," *International Journal of Production Research*, vol. 9, no. 4, 1971.
23. Buzacott, J. A., "Optimal Operating Rules for Automated Manufacturing Systems," *IEEE Transactions on Automatic Control*, vol. 27, no. 1, February 1982.
24. Buzacott, J. A., "The Fundamental Principles of Flexibility in Manufacturing Systems," Proceedings of the International Conference on Flexible Manufacturing Systems, October 1982.
25. Buzacott, J. A., "A Perspective On New Paradigms In Manufacturing," *Journal of Manufacturing Systems*, vol. 14, no. 2, 1995.
26. Buzacott, J. A., and M. Mandelbaum, "Flexibility and Productivity in Manufacturing Systems," Annual International Industrial Engineering Conference Proceedings, 1985.
27. Buzacott, J. A., and J. G. Shanthikumar, "Models for Understanding Flexible Manufacturing Systems," *AIIE Transactions*, December 1980.
28. Carroll, D. C., "Heuristic Sequencing of Single and Multiple Component Jobs," Ph. D. dissertation, Sloan School of Management, M.I.T., 1965.
29. Chambers, J. C., S. K. Mullick, and D. D. Smith, "How To Choose the Right Forecasting Technique," *Harvard Business Review*, July–August 1971.
30. Clark, J. T., "Selling Top Management—Understanding the Financial Impact of Manufacturing Systems," APICS Conference Proceedings, 1982.
31. Conway, R. W., W. L. Maxwell, and L. W. Miller, *Theory of Scheduling*, Addison-Wesley, 1967.
32. Crosby, P. B., *Quality Is Free*, McGraw-Hill, 1979.
33. Cross, M., *Managing Workforce Reduction*, Praeger, 1985.
34. Dauch, R. E., *Passion for Manufacturing*, Society of Manufacturing Engineers, 1993.
35. Davis, B. L., L. W. Hellervik, C. J. Skube, S. H. Gebelein, and J. L., Sheard, *Successful Manager's Handbook*, Personnel Decisions, 1992.
36. Deal, T. E., and A. A., Kennedy, *Corporate Cultures*, Addison-Wesley, 1982.
37. Deming, E. W., *Out of the Crisis*, Massachusetts Institute of Technology, 1982.
38. Deming, E. W., *Quality, Productivity, and Competitive Position*, Massachusetts Institute of Technology, 1982.
39. Drucker, P. F., *Management*, Harper and Row, 1974.
40. Eccles, R. G., and N. Nohria, *Beyond the Hype*, Harvard Business School Press, 1992.

41. Eureka, W. E., and N. E. Ryan, *The Customer Driven Company*, ASI Press, 1988.
42. Fogarty, D. W., J. H. Blackstone, and T. R. Hoffmann, *Production and Inventory Management*, 2d ed., South-Western, 1991.
43. Ford, H., *Today and Tomorrow*, Productivity Press, 1988.
44. Gardiner, S. C., J. H. Blackstone, and L. R. Gardiner, "The Evolution of the Theory of Constraints," *Industrial Management*, May–June 1994.
45. Gargeya, V. B., and J. P. Thompson, "Just-In-Time Production in Small Job Shops," *Industrial Management*, July–August 1994.
46. Gass, S. I., *Linear Programming: Methods and Applications*, McGraw-Hill, 1985.
47. Goldman, S. L., R. N. Nagel, and K. Preiss, *Agile Competitors and Virtual Organizations*, Van Norstrand Reinhold, 1995.
48. Goldratt, E. M., and J. Cox, *The Goal*, North River Press, 1984.
49. Goldratt, E. M., and R. E. Fox, *The Race*, North River Press, 1986.
50. Greenberg, H., "A Branch and Bound Solution to the General Scheduling Problem," *Operations Research*, vol. 16, no. 2, March 1968.
51. Greene, J. H., *Production and Inventory Control Handbook*, McGraw-Hill, 1987.
52. Gross, D., and C. Harris, *Fundamentals of Queueing Theory*, John Wiley and Sons, 1974.
53. Gue, F. M., *Increased Profits Through Better Control of Work In Process*, Reston, 1980.
54. Gupta, J. N., and R. A. Dudek, "Optimality Criteria for Flow Shop Schedules," *AIIE Transactions*, vol. 3, no. 3, September 1971.
55. Hammer, M., and J. Champy, *Reengineering the Corporation*, HarperCollins, 1993.
56. Hammer, M., and S. A. Stanton, *The Reengineering Revolution*, HarperCollins, 1995.
57. Hariharan, R., and P. Zipkin, "Customer-Order Information, Lead times, and Inventories," Management Science, vol. 41, no. 10, October 1995.
58. Harmon, R. L. and L. D. Peterson, *Reinventing the Factory*, Andersen Consulting, 1990.
59. Hartley, R. V., "Transmission of Information," *Bell Systems Technical Journal*, no. 535, 1928.
60. Hay, E. J., *The Just-In-Time Breakthrough*, John Wiley and Sons, 1988.
61. Hayes, R. H., and S. C. Wheelwright, *Restoring Our Competitive Edge*, John Wiley and Sons, 1984.
62. Hayes, R. H., and S. C. Wheelwright, *Dynamic Manufacturing*, Free Press, 1988.

63. Hendricks, K. B., and J. O. McClain, "The Output of Serial Production Lines of General Machines with Finite Buffers," *Management Science*, vol. 39, no. 10, October 1993.

64. Hersey, P., and K. H. Blanchard, *Management of Organizational Behavior*, Prentice-Hall, 1977.

65. Hickman, C. R., and M. A. Silva, *Creating Excellence*, Plume, 1984.

66. Hill, T., *Manufacturing Strategy: Text and Cases*, Irwin, 1989.

67. Hillier, F. S., and G. J. Lieberman, *Introduction to Operations Research*, Holden Day, 1974.

68. Hitt, W. D., *Ethics and Leadership*, Battelle Press, 1990.

69. Hodgetts, R. M., *Management*, Academic Press, 1985.

70. Honeycutt, E. D., J. A. Siguaw, and S. C. Harper, "The Impact of Flexible Manufacturing On Competitive Strategy," *Industrial Management*, November–December 1993.

71. Hu, T. C., "Parallel Sequencing and Assembly Line Problems," *Operations Research*, vol. 9, no. 6, November 1961.

72. Iman, R. L., and W. J. Conover, *Modern Business Statistics*, John Wiley and Sons, 1983.

73. Ishikawa, K., *Guide to Quality Control*, Asian Productivity Organization, 1976.

74. Jalinek, M., and J. D. Goldhar, "The Strategic Implications of the Factory of the Future," *Sloan Management Review*, Summer 1984.

75. Johnson, S. M., "Optimal Two and Three-Stage Production Schedules with Setup Times Included," *Naval Research Logistics Quarterly*, vol. 1, no. 1, March 1954.

76. Johnson, T., and R. S. Kaplan, *Relevance Lost*, Harvard Business School Press, 1987.

77. Juran, J. M., and F. M. Gryna, *Juran's Quality Control Handbook*, 4th ed., McGraw-Hill, 1988.

78. Kanter, R. M., *The Change Masters*, Simon and Schuster, 1983.

79. Kaplan, R. S., *Measures of Manufacturing Excellence*, Harvard Business School Press, 1990.

80. Karger, D. W., and F. H. Bayha, *Engineered Work Measurement*, Industrial Press, 1977.

81. Kautz, W. H., and M. Branon, *Intuiting the Future*, Harper and Row, 1989.

82. Knott, A. D., "The Inefficiency of a Series of Work Stations—A Simple Formula," *International Journal of Production Research*, vol. 8, no. 109, 1969.

83. Kotter, J. P., and J. L. Heskett, *Corporate Culture and Performance*, Free Press, 1992.

84. LaBarre, P., "The New Strategic Paradigm," *Industry Week*, November 1994.

85. Leeds, D., *Smart Questions*, McGraw-Hill, 1987.

86. Levitt, T., *The Marketing Imagination*, Free Press, 1986.

87. Lomnicki, Z., "A Branch-and-Bound Algorithm for the Exact Solution of the Three-Machine Scheduling Problem," *Operations Research Quarterly*, vol. 16, no. 1, March 1965.

88. Lu, D. J., *Kanban*, Productivity Press, 1985.

89. Mahoney, R. M., "TQC For Manual Processes in an Electronics Manufacturing Environment," Proceedings ATE&I Instrumentation Conference East, 1989

90. Mahoney, R. M., and G. L. Larsen, "Economic Analysis Clarifies Costs of Final Test and Board Test," *Electronics Test*, November 1988.

91. Mahoney, R. M., and G. L. Larsen, "A Bayesian Approach to Improving Test Effectiveness and Evaluating Test Strategy," Proceedings ATE&I Instrumentation Conference West, 1990.

92. Malcolm, J. G., and G. L. Foreman, "The Need: Improved Diagnostics—Rather than Improved R," Proceedings Annual Reliability and Maintainability Symposium, 1984, pp. 315–322.

93. Mayo, E., *The Human Problems of an Industrial Civilization*, Macmillan, 1933.

94. McNaughton, R., "Scheduling with Deadlines and Loss Functions," *Management Science*, vol. 6, no. 1, October 1959.

95. Mitroff, I. I., *Business Not As Usual*, Jossey-Bass, 1987.

96. Mitten, L. G., "Branch and Bound Methods: General Formulation and Properties," *Operations Research*, vol. 18, no. 1, January–February 1970.

97. Mondon, Y., *Toyota Production System*, Institute of Industrial Engineers, 1983.

98. Moody, P. E., *Strategic Manufacturing*, Irwin, 1990.

99. Moore, C. L., and R. K. Jaedicke, *Managerial Accounting*, South-Western, 1980.

100. Newman, W. R., and M. D. Hanna, "Including Equipment Flexibility in Break-even Analysis: Two Examples," *Production and Inventory Management Journal*, 1st Quarter 1994.

101. Nolan, R. l., and D. C. Croson, *Creative Destruction*, Harvard Business School Press, 1995.

102. Oden, H. W., G. A. Langenwalter, and R. A. Lucier, *Handbook of Material and Capacity Requirements Planning*, McGraw-Hill, 1993.

103. O'Guin, M. C., *Activity Based Costing*, Prentice-Hall, 1991.

104. Ohmae, K., *The Mind of the Strategist*, McGraw-Hill, 1982.

105. Ott, E. R., *Process Quality Control*, McGraw-Hill, 1975.

106. Packard, D., *The HP Way*, HarperCollins, 1995.

107. Park, J., M. Kang, and K. Lee, "An Intelligent Operations Scheduling

System in a Job Shop," *International Journal of Advanced Manufacturing*, 1966.

108. Park, P. S., "Uniform Plant Loading Through Level Production," *Production and Inventory Management Journal*, 2d Quarter 1993.

109. Pascale, R. T., *Managing on the Edge*, Simon and Schuster, 1990.

110. Pasmore, W. A., *Designing Effective Organizations*, John Wiley and Sons, 1988.

111. Patterson, M. L., *Accelerating Innovation*, Van Norstrand Reinhold, 1993.

112. Peters, T., *Thriving On Chaos*, Alfred A. Knopf, 1987.

113. Pfeffer, J., *Managing with Power*, Harvard Business School Press, 1992.

114. Plossl, G. W., *Manufacturing Control: The Last Frontier for Profits*, Reston, 1973.

115. Plossl, G. W., *Production and Inventory Control: Principles and Techniques*, Prentice-Hall, 1985.

116. Plossl, G. W., *Engineering for the Control of Manufacturing*, Prentice-Hall, 1987.

117. Plossl, G. W., *Managing in the New World of Manufacturing*, Prentice-Hall, 1991.

118. Plossl, G. W., *Orlicky's Material Requirements Planning*, McGraw-Hill, 1994.

119. Proud, J. F., *Master Scheduling*, John Wiley and Sons, 1994.

120. Pugh, S., *Total Design*, Addison-Wesley, 1991.

121. Raturi, A. S., J. R. Meredith, D. M. McCutcheon, and J. D. Camm, "Coping with the Build-to-Forecast Environment," *Journal of Operations Management*, vol. 9, no. 2, April 1990.

122. Roberts, G. W., *Quality Assurance in Research and Development*, Marcel Dekker, 1983.

123. Ronen B., and Y. Spector, "Managing System Constraints: A Cost/Utilization Approach," *International Journal of Production Research*, vol. 30, no. 9, 1992.

124. Rummler, G. A., and A. P. Brache, *Improving Performance*, Jossey-Bass, 1990.

125. Sandras, W. A., *Just-In-Time: Making It Happen*, Oliver Wight Publications, 1989.

126. Schiller, B. R., *The Micro Economy Today*, Random House, 1989.

127. Schmenner, R. W., "The Merit of Making Things Fast," *Sloan Management Review*, Fall 1988.

128. Schonberger, R. J., *Japanese Manufacturing Techniques*, Free Press, 1982.

129. Schrage, L. E., "A Proof of the Optimality of the Shortest Remaining Processing Time," *Operations Research*, vol. 6, no. 3, May–June 1968.

130. Sell, P. S., *Expert Systems: A Practical Introduction*, John Wiley and Sons, 1985.

131. Shanklin, W. L., and J. K. Ryans, *Thinking Strategically*, Random House, 1985.

132. Sharma, K., "Adding Intelligence to MRP Systems," *APICS—The Performance Advantage*, March 1993.

133. Shingo, S., *Zero Quality Control: Source Inspection and the Poka-Yoke System*, Productivity Press, 1986.

134. Shingo, S., *Study of Toyota Production System from Industrial Engineering Viewpoint*, Japan Management Association, 1981.

135. Shores, R. A., *Survival of the Fittest*, ASQC Quality Press, 1988.

136. Skinner, W., "The Focused Factory," *Harvard Business Review*, May–June 1974.

137. Smith, B. T., "Focus Forecasting: A New Level of Accuracy in Inventory Management," *Modern Materials Handling*, September 1978.

138. Smith, B. T., *Focus Forecasting*, Oliver Wight Publications, 1984.

139. Smith, M. R., *Bottom-Line Plant Management*, Prentice-Hall, *1991*.

140. Smith, R. D., and R. A. Dudek, "A General Algorithm for Solution of the n-Job m-Machine Sequencing Problem of the Flow Shop," *Operations Research*, vol. 15, no. 1, January 1967.

141. Srikanth, M. L., and S. A. Robertson, *Measurements for Effective Decision Making*, Spectrum, 1995.

142. Stalk, G., "Time—The Next Source of Competitive Advantage," *Harvard Business Review*, July–August 1988.

143. Stalk, G., and T. M. Hout, *Competing Against Time*, Free Press, 1990.

144. Stamper, D. A., *Business Data Communications*, Benjamin/Cummings, 1986.

145. Steiner, G., and S. Yeomans, "Level Schedules for Mixed-Model Just-In-Time Processes," *Management Science*, vol. 39, no. 6, June 1993.

146. Suarez, F. F., "An Empirical Study of Manufacturing Flexibility in Printed-Circuit Board Assembly," Working Paper, Sloan School of Management, M.I.T., 1992.

147. Sumanth, D. J., "Productivity Management—A Challenge for the 80s," ASEM Proceedings, First Annual Conference, American Society for Engineering Management, 1980.

148. Sumanth, D. J., *Productivity Engineering and Management*, McGraw-Hill, 1984.

149. Sumanth, D. J., *Productivity Management Frontiers–I*, Elsevier, 1987.

150. Szendrovits, A. Z., "Manufacturing Cycle Time Determination for a Multi-Stage Economic Production Quantity Model," *Management Science*, vol. 22, no. 3, 1975.

151. Taylor, F. W., *Shop Management*, Harper and Brothers, 1919.
152. Taylor, F. W., *The Principles of Scientific Management*, Hive, 1911, 1985.
153. Thurow, L. C., *The Zero Sum Solution*, Simon and Schuster, 1985.
154. Tregoe, B. B., J. W. Zimmerman, R. A. Smith, and P. M. Tobia, *Vision In Action*, Simon and Schuster, 1989.
155. Truscott, W. G., "Scheduling Production Activities in Multi-Stage Batch Manufacturing Systems," *International Journal of Production Research*, vol. 23, no. 2, 1985.
156. Turban, E., *Decision Support and Expert Systems*, Macmillan, 1988.
157. Umble, M., and M. L. Srikanth, *Synchronous Manufacturing*, Spectrum, 1995.
158. Upton, D. M., "The Management of Manufacturing Flexibility," *California Management Review*, Winter 1994.
159. Vollmann, T. E., W. L. Berry, and C. D. Whybark, *Manufacturing Planning and Control Systems*, Irwin, 1988.
160. Wagner, H. M., *Principles of Operations Research*, Prentice-Hall, 1975.
161. Walton, M., *The Deming Management Method*, Dodd Mead, 1986.
162. Wantuck, K. A., *Just In Time for America*, KWA Media, 1989.
163. White, J. A., *Production Handbook*, John Wiley and Sons, 1987.
164. Wight, O. W., *Production and Inventory Management in the Computer Age*, Cahners, 1974.
165. Wild, R., *International Handbook of Production and Operations Management*, Cassell, 1989.
166. Zweben, M., and M. S. Fox, *Intelligent Scheduling*, Morgan Kaufmann, 1994.

Index

A

ABC (activity based costing) *145, 148, 149*
ABM (activity based management) *145*
Action Bucket *56*
Activation *43*
Agile manufacturing *7, 8, 90*
Allocation *64, 69*
Andon *34*
ANSI X-12 standard *66*
APICS (American Production and Inventory Control Society) *194*
Assemble-to-order *49, 51, 52*
Asset utilization *123, 163, 164, 200*
ATP (available-to-promise) *57, 58*
Automation *71, 125, 150, 187*
Automotive (OEMs) *4*
Availability *43*
Available capacity *28, 43, 65, 71, 80, 117, 123, 158, 160*

B

Backward-scheduling *64, 65, 71, 80, 170, 175*
Balance delay *120, 174, 200*
Balance sheet *107, 143, 188, 195*
Baldrige, Malcom *3*
Baseline measure *199, 200*
Benchmark *6, 120, 188, 206*
Bill of resources *49, 59*
Black hole *175*
Blocking *35, 39, 40*
BOM (bill of material) *60, 63, 173*
Book value (machine) *200, 203*
Bottleneck *37–39, 78, 90, 122, 123, 150, 196, 199, 200*
Branch and bound *81, 124*
BSS (business strategy statement) *46*
Bucketless *60*
Buffer inventory *17, 35, 59, 180–184*
Built-to-order *44, 45, 49, 51, 52, 101, 115*
Build-to-stock *44, 45, 49, 51, 52, 101, 115*
Bureaucrat *156*
Business plan *46, 47, 49, 69, 206*
Buyers *53, 66*

C

CALB (computer-aided line balancing) *124, 125*
Capacity *42, 43, 90, 157–160*
Capacity buffer *42, 80*
Carroll *84*
Cash flow *107*
Ceiling *121, 122, 124, 170*
Chess *9*
Class-A MRP II *206*
Coefficient of variation *85*
Competitive advantage *1*
Competitiveness *1, 8, 10*
Complacency *207*
Complexity *6, 7, 95, 129, 138, 145*
Comptroller *206*
Concept value delivery chain *19, 20*
Conflict *188–190*
Constraint *34–43, 57, 71, 78, 116, 123, 150*
Continuous improvement model *141*
Continuous quality improvement *67, 131, 186*
Controller *53, 206*
Conversion efficiency *202, 203*
Cosourcing *127, 160, 170*
Cost accounting *16, 85, 143–145, 191, 200, 203*
Cost accumulation factor *146, 147*
Cost cutting *185–187*
Costing *143*
CPIM (certification in production and inventory management) *194*
CRP (capacity requirements planning) *64, 65*
Culture *132, 141, 156, 188, 189*
Customer *1, 3, 7, 8, 34, 50, 51, 59, 80, 85, 115, 193*
CVBA (cost-volume, break-even analysis) *125, 126*
Cycle counting *64*
Cycle time *58, 69, 79, 122–124, 139, 207*

D

Decision support system *176*
Decision theory *155*

217